Praise for *The Ever*

As an ever-curious gardener who seeks to understand the science behind all things gardening, I look to my horticultural heroes for that. Lee Reich is always one of my top go-to authorities. Much to my delight, and no surprise, this book is everything I was hoping and more. Leave it to Lee to blend science with real-world application, mixed with a chuckle or two throughout the pages. A fresh, fun, and fascinating must-read for every curious gardener.

—Joe Lamp'l, Creator & Host, PBS's Growing a Greener World

Behind the pleasures of the successful garden, there are the apparent mysteries. How does it all work? Curious gardeners have questions and Lee Reich answers them as effectively as that favorite science teacher in school did—clearly and concisely.

—Eliot Coleman, farmer; past Executive Director, International Federation of Organic Agriculture Movements; host, "Gardening Naturally;" and author, *Four Season Harvest*

Armed with Lee Reich's brand of gardening science, I brush a hand along the tops of my seedlings so they grow sturdier stems. Or I jostle their trays and say, "Good morning." When hoeing weeds, I wear a long skirt that brushes the tops of my cucumber and melon plants and they produce more female flowers, hence, more fruit. Combining scientific reasoning with the joy of touch and observation will not only make you a better gardener, you'll find yourself with permission to play in a curious world full of intrigue and creativity.

—MaryJane Butters, organic farmer, small dairy owner, beekeeper, author, magazine editor MaryJanesFarm.org

As a commercial grower, I don't read that many gardening books... although I make an exception for Lee's Reich books. *The Ever Curious Gardener* explains some of the science behind what's going on above ground and below ground in your garden and—most important—how you can work with these natural systems to grow plants that are healthier, more productive, and more attractive. For a better garden and more interesting gardening, read this book.

—Jean-Martin Fortier, author, *The Market Gardener*

Perhaps the most readable gardening book that I have ever read. Full of carefully presented garden practices that are supported by scientific know-how, it's fun and informative. What more can I say?

—Jeffrey Gillman, author, *The Truth About Garden Remedies*

The Ever Curious Gardener is a wise and witty book that offers not just the "how-to" of gardening but also the "how come?" By showing you the science behind growing plants, it gives you the tools to follow Nature's rules—the only ones that count. What Harold McGee is to the cook, Lee Reich is to the gardener.

—Barbara Damrosch, author, *The Garden Primer* and *The Four Season Farm Gardener's Cookbook*

Gardeners in the know wait for books by Lee Reich, and *The Ever Curious Gardener* shows why. Very few writers combine science, history, and personal observations to produce a great (and humorous) read with so much practical advice quite like Reich. This latest, and hopefully not last book, does not disappoint!

—Jeff Lowenfels, author, *Teaming with Microbes* series

From root to shoot, flower to fruit, here is an essential field guide to the science behind plant cultivation. Chapter by chapter it will steer the gardener's hand and delight the mind at the same time.

—Roger B. Swain, Host, PBS-TV's "The Victory Garden"

There are far too few garden scientists with the ability to write for a popular audience, so thank goodness Lee Reich can do just that! His newest book is full of current, factual information that's of immediate use to gardeners everywhere. It's the perfect excuse to "let your curiosity get the better of you."

—Dr. Linda Chalker-Scott, WSU horticulturist and author, *The Informed Gardener* series and *How Plants Work: The Science Behind the Amazing Things Plants Do*

With *The Ever Curious Gardener*, Lee Reich presents some of the natural science behind the scenes in the garden. Not in a detached, academic manner, but pragmatically (and sometimes humorously), as it can be applied to make for a better garden and gardener. Read it and reap.

—Ron Khosla, Professor Environmental Sciences, Southern Oregon University, International Consultant to United Nations FAO, and founder, Huguenot Street Farm

The Ever Curious Gardener cultivates curiosity and brings out everyone's inner science nerd. Lee Reich's engaging and authentic style blend science with practical gardening knowledge. Anyone reading these pages is guaranteed to harvest new, insightful knowledge.

—Lisa Kivirist, author, *Soil Sisters: A Toolkit for Women Farmers* and *Homemade for Sale*

THE *Ever Curious* GARDENER

USING A LITTLE NATURAL SCIENCE FOR A MUCH BETTER GARDEN

by

Lee Reich

Illustrations by Vicki Herzfeld Arlein

Cover design by Diane McIntosh.
Cover images: © iStock (mag. glass: 155383741, label paper texture: 483534560, plant diagram: 507214734, sprout: 511977848, plant cells: 578118802, bee on squash blossom: 599702254, squash plant: 813245542.jpg)
All interior photos © Lee Reich unless otherwise credited.

Printed in Canada. First printing March 2018.

Inquiries regarding requests to reprint all or part of *The Ever Curious Gardener* should be addressed to New Society Publishers at the address below. To order directly from the publishers, please call toll-free (North America) 1-800-567-6772, or order online at www.newsociety.com

Any other inquiries can be directed by mail to:

New Society Publishers
P.O. Box 189, Gabriola Island, BC V0R 1X0, Canada
(250) 247-9737

LIBRARY AND ARCHIVES CANADA CATALOGUING IN PUBLICATION

Reich, Lee, author
The ever curious gardener : using a little natural science for a much better garden / by Lee Reich ; illustrations by Vicki Herzfeld Arlein.

Includes index.
Issued in print and electronic formats.
ISBN 978-0-86571-882-1 (softcover).—ISBN 978-1-55092-675-0 (PDF).—
ISBN 978-1-77142-270-3 (EPUB)

1. Gardening. I. Herzfeld Arlein, Vicki, illustrator II. Title.

SB450.97.R45 2018 635 C2018-900123-2
C2018-900124-0

New Society Publishers' mission is to publish books that contribute in fundamental ways to building an ecologically sustainable and just society, and to do so with the least possible impact on the environment, in a manner that models this vision.

DEDICATION

To my father, Joseph Reich,

who early on encouraged me to be

curiouser and *curiouser*.

Contents

Acknowledgments

A number of people helped bring this book to fruition. Helpful comments, corrections, and suggestions were offered by Bob Arlein, Janna Beckerman, Sara Gast, Genevieve Reich, Drew Waddelow, Deb Goldman, and David Weisberger. The idea for the one sentence summaries, à la Tom Jones, that introduce each chapter's subsections was sparked by Peter Mayer. Special thanks go to Vicki Herzfeld Arlein for her many insightful comments on text and design, as well as for her artful and lucid illustrations.

I feel especially lucky to have worked with New Society Publishers. Thanks to Rob West and Ingrid Witvoet for getting the ball rolling, Sue Custance for keeping the ball rolling, Greg Green for the design process, and EJ Hurst for marketing. I appreciate editor Ian le Cheminant's untiring attention to detail as well as his technical and artistic command of the language.

Introduction

"Though an old man, I am a young gardener." So wrote Thomas Jefferson. The longer I garden, the more I realize the truth of those words. Gardening is a lifelong learning experience that never ceases to capture, recapture, and then capture my interest once again. How could it not, representing, as it does, such a congenial confluence of colors, flavors, and aromas all seasoned with the weather, whatever pests happen to stop by that year—and the science behind it all?!

And the science behind it all is what this book is about. It's not a comprehensive overview of botany and related sciences, just some natural science that can be applied in the garden. No need to read from cover to cover or in one fell swoop to get the most out of this book. Each chapter can stand by itself—as, in most cases, can each section within a chapter. So dip in and out of this book according to your whim, the season, or what's happening in your garden.

Science may seem out of place in so bucolic an activity as gardening. After all, millions of years of evolution have prompted seeds to germinate and plants to grow in soils and climates as diverse as the Arctic tundra, the Arizona desert, and my garden in New York's Hudson Valley. So it's possible to have a decent garden with minimal effort or know-how.

But gardening can be something more than this business as usual, with commensurately more rewards.

* * *

The genesis for this book came to me one day as I was piling scythed meadow hay and horse manure, along with old vegetable plants and sprinklings of soil and dolomitic limestone, into one of my compost bins. I realized that what I was adding to the pile and how much of each ingredient, even how I fluffed them up or patted them down with my pitchfork, and then watered, all reflected what I had learned over the past 40-plus years of gardening. My classrooms have included actual classrooms; gleanings from magazines, books, and scientific journals; conversations with other gardeners and agricultural scientists; and (most importantly) the garden itself.

My garden education has been unusual. Growing up in the suburbs, I initially remembered only a small vegetable garden whose tenure was soon eclipsed by a swing set. Wait! How about that potted banana tree and one hyacinth bulb that I nurtured under the purple glow of a Growlite in the basement during high school? Or the potted cactus that I bought to adorn my bedroom windowsill in graduate school. Hints of future interest? Perhaps.

Graduate study in those cactus days was in chemistry, a continuation of an interest kindled by my high school chemistry teacher. But coming to the conclusion that graduate study in quantum chemistry was not going to answer any fundamental questions, I dropped out, moved to Vermont, and got the gardening bug. Because I was living in a third floor apartment, I expressed that gardening bug with a voracious appetite for books—books about gardening.

A year later, I dove into agriculture in earnest, and was fortunate to land in a graduate program in soil science. My interest and education in chemistry proved a good foundation for soil science.

A small plot of land began my education "in the field" and complemented my academic studies. The university's agricultural library offered more books to further round out my education. (I remember coming across a whole book on lettuce seed!)

Eight years later, with two framed diplomas to hang on my wall, one for a master's degree in soil science, and the other for a doctorate degree in horticulture, I was still gardening with the same exuberance and learning about gardening through experience, the printed word, and contact with others "in the know." Thinking back, how little I knew about gardening. And so it goes.

* * *

Back to my compost pile...I took into account the meadow hay's youthful lushness, which influences its ratio of carbon to nitrogen, as I layered it into the bin along with the horse manure. Manure is usually thought of as a high nitrogen material, but I looked at what was in the cart and, eyeing the amount and kind of bedding (wood shavings) with which it was mixed, made a rough estimate in my head of how much to use to make a good balance with the meadow plants. When the pile was finished, I checked my work by monitoring the temperature of the pile's interior with a long-stemmed compost thermometer. Etc., etc. There's art in making compost. But also science.

With this book, I hope to show you, the reader, how knowing and using a little of the natural science behind what's happening out in the garden can make for a lot better garden in terms of productivity, beauty, plant health, sustainability...and interest. Knowing some of the underlying science at work in the garden also makes for a more resilient gardener, better able to garden at a new location or in a changing environment. All of which makes for a perennially "young gardener," as Mr. Jefferson said it!

PROPAGATION AND PLANTING

A bit of deception helps me get some seeds to sprout that under natural conditions would wisely stay asleep.

You wouldn't think that the dead of winter would be a good time to sow seeds. But it is, for plants whose seeds need some kind of long term treatment before they will sprout. Such is the case for the tree peony seeds I recently planted.

"Planted" is really too gardenesque a term for what I did with those seeds. After soaking them in water for a few hours, I merely tossed them into a plastic sandwich bag with a handful of moist potting soil. The bag will sit on the kitchen counter for a couple of months, then go into the refrigerator for another couple of months.

Peony seeds need this treatment because they must lay down roots before any shoot growth can begin. To grow roots, those seeds need some rain (or a good soaking) to leach inhibitors from the seeds, and they also need some warmth. The shoots, however, won't sprout until they've been exposed to a period of cool, moist conditions—outdoors or in my refrigerator. Under natural conditions, all this might take two years. In my house, all systems should be go by spring. Lily and viburnum seeds also respond to this type of treatment.

A reluctance to sprout as soon as touching down on moist soil often makes sense for ensuring the survival of tender, young seedlings. Not rearing their heads until convinced that winter is over and they have the support of an established root system is just the ticket for survival of wild tree peony seedlings in a climate characterized by cold winters and periodic drought.

Germination quirks of other kinds of plant seeds reflect other natural environments. Some seeds have a double dormancy, one for the seed coat and one for the embryo. Still others (goldenseal, for example) ripen with underdeveloped embryos. The same warm, then cold, treatment needed by tree peonies also prepares seeds with either of these quirks for germination.

Where moisture is more or less consistent throughout the year, it is winter cold that would kill a young sprout that began growing in the fall. Fall-ripening seeds won't sprout until they feel that winter is over, a condition that could be mimicked by a couple of months in the refrigerator in a sandwich bag along with moist potting soil. After doing time in the refrigerator, it's not unusual for a whole batch of seeds to sprout in unison, as if a switch has been turned on, even before they're released into warmth. That cool, moist treatment is called stratification because in the past nurseries effected this treatment by spreading alternating layers of seeds and soil in flats kept outdoors for the winter.

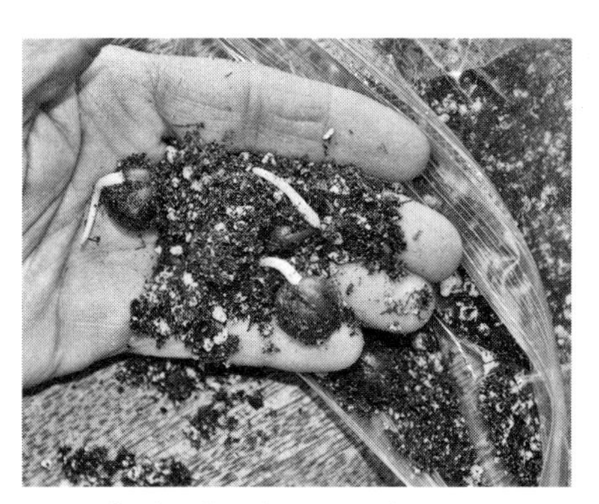

Stratified yellowhorn seeds sprouting.

Hormones within seeds are what bring them to life at the appropriate moment. Although lying apparently lifeless in a bag on a refrigerator shelf, all sorts of things—hormonally—are going on. Levels, for instance, of a germination inhibitor called abscisic acid are decreasing while levels of another hormone, gibberellic acid, are increasing. These hormones have been extracted from seeds and synthesized. Some seeds shed their normal reluctance to sprout with nothing more than a dip in an appropriate concentration of gibberellic acid. All is not so simple, though, because other hormones are also at work, and other compounds, such as potassium nitrate, hydrogen peroxide, or malt extract can also promote germination.

Not all fall-ripening seeds need stratification before they will germinate. Two examples of tree seeds in this class are those of catalpa and those of sycamore (although sycamore's relative, the London planetree, does need stratification). Perhaps catalpa and sycamore seeds have evolved without a need for stratification because they hang on the trees late enough into the winter so that, by the time they drop to moist ground, temperatures are too cold for germination. Or else spring has arrived, and it's just the right time for germination.

Let's not blame dormancy only on hormones; some seeds stay asleep for purely mechanical reasons. The tough seed coats of honeylocust, black locust, and black cohosh are among those that can't imbibe water as soon as their seeds hit the ground. A seed that remains too dry inside will not sprout. These are examples of seeds that need scarification before they can be stratified.

In nature, tough coats are eventually softened—as soil microbes chew away at them, by cycles of freezing and thawing, by abrasion, and by passage through animals. Microbes work best at warm temperatures, so a couple of months in a sandwich bag along with some potting soil could awaken these seeds just as they do those of tree peonies. The potting soil, in this case, should contain some real soil or compost to supply living organisms to work on the seed coats.

Scarification means "to scratch" and with large enough seeds I take this meaning literally, with a file. Nicking seeds or nipping out a little piece of seed coat with a wire cutter are other ways to let water in past a tough seed coat. A quicker way to scarify a batch of seeds is with very hot water or even sulfuric acid, but care is needed not to kill the seeds. As a general rule, bring almost to a boil 5 times the volume of water as the volume of seeds, then pour

Scarifying hard yellowhorn seeds with a file.

the water over the seeds and let them stand in the water for 12 to 24 hours. With sulfuric acid, suffice it to say that familiarity with using this caustic chemical is needed, along with goggles and gloves. Timing is critical, and varies with the kind of seed. The acid must be thoroughly rinsed off following the treatment.

The easiest pretreatment is that needed by many grasses and most annual flowers and vegetables. Seeds of these plants need nothing more than a period of dry storage of from one to six months before they'll germinate. Cold is not needed, but does keep them fresh longer—so my vegetable and flower seeds wait out winter sitting in airtight plastic boxes and Mason jars in my garage.

Burial in tundra might be ideal for seed storage but I choose more practical storage for my vegetable and flower seeds.

Few seeds have as short a viability as onion; after only a year they might not be sufficiently viable for sowing.

A better story is the reported longevity of the 10,000 year old lupine seed that germinated after being taken out of a lemming burrow in the Yukon permafrost. Just think: the plant that produced this seed was up and growing when humans first walked across the Bering Land Bridge, and saber-toothed cats and woolly mammoths may have brushed up against its leaves. Except that the story of the 10,000 year old lupine seed turns out to be apocryphal, as confirmed by radiocarbon dating.

The true record for seed longevity was, until recently, 2,000 years, and was held by a date palm grown from seed recovered from an ancient fortress in Israel. A more recent discovery broke that record by a long shot.

A kind of campion seed (*Silene stenophylla*) found buried, this time in a squirrel burrow in Siberian tundra, could very well be 32,000 years old. The seed sprouted and was grown into a charming, white-flowered plant. Some coaxing was needed to get that seed to sprout. Actually, the seed itself did not sprout, but new plants were propagated from a few cells that were removed from the placenta and multiplied under sterile conditions on a specially concocted growth medium. Once cells had multiplied sufficiently, the growing medium was altered to induce growth of leaves, stems, and roots, and eventually the plants were robust enough to be planted in soil. The plant flowered and set seed, which germinated readily to produce more seedlings.

As short as is onion seed viability (I purchase new seed every year), other seeds have even shorter viability. Seeds of some members of the subfamily Tillandsioideae, related to pineapple, remain viable for only 4 to 6 weeks. Silver maple, *Acer saccharinum*, seeds retain their capacity to germinate for only about a week, making the many silver maples in the view out my bedroom window testimonial to the trees' fecundity.

Viable seeds are living, albeit dormant, embryonic plants which do not live forever. It's wasted effort to sprinkle dead seeds into furrows either in the garden or seed flats.

When purchasing a packet of seeds from a local store or mail-order seedhouse, you are assured of the viability of the seeds. There are government standards for the minimum percentage of seeds that must germinate for each type of seed. The packing date and the germination percentage often are stamped on the packets. (The germination percentage must be indicated only if it is below standard.) I write the year on any seed packets on which the date is not stamped.

Old, dog-eared seed packets may or may not be worth using this season. It depends on where the packets were kept and the types of seeds they contain.

Conditions that slow biological and chemical reactions, such as low temperature, low humidity, and low oxygen, also slow the aging of seeds. During spring and summer, the airtight plastic boxes and Mason jars in which I store seeds find their low temperature and low humidity home in the depths of my freezer or, more recently, in the cool temperatures of my basement. By fall, when frozen fruits and vegetables claim freezer space, I move seed boxes and jars back to the garage. An easy way to keep the humidity low in the storage containers is with silica gel or by sprinkling in some powdered milk, from a freshly opened box. Renew the powdered milk each year. Silica gel can be renewed in a hot oven.

It's not impossible for a backyard gardener to store seeds in a low oxygen atmosphere. I reverse engineered a bicycle pump to become a weak vacuum pump which, along with a Foodsaver® Wide Mouth Jar Sealer Vacuum Sealing Accessory, evacuates some of the air from my seed-containing Mason jars. Thinner air is also drier air.

The air in a sealed jar could, instead, be displaced before sealing it with a gas other than oxygen. Carbon dioxide is readily available in cartridges; a carbon dioxide bicycle tire inflator could be used to direct this gas into the jar. Argon gas is another option (Bloxygen®) that's used to preserve various products. Important when using either of these products is to introduce the gas slowly to avoid turbulence and allow it to settle. Both gases are heavier than oxygen.

Although some seed companies market their seeds in hermetically sealed, plastic-lined foil packets, I've never noted superior germination from these foil packets, as compared with plain old paper packets. Matter of fact, my own casual observations over the years are that germination of seeds kept in these hermetically sealed packets is worse. Perhaps the extra cost of the packaging is a disincentive to a seedhouse to discard old seeds or open the packets for re-testing. Perhaps my casual observations are too casual.

Seeds differ in how long they remain viable. Except under the very best storage conditions, as with onion seed, it's not worth the

Vegetable Seed Longevity Under Good Storage Conditions

Vegetable	Years	Vegetable	Years	Vegetable	Years
Bean	3	Cucumber	5	Parsnip	1
Beet	4	Eggplant	4	Pea	3
Broccoli	3	Endive	5	Pepper	2
Brussels sprouts	4	Fennel	4	Pumpkin	4
Cabbage	4	Kale	4	Radish	5
Carrot	3	Kohlrabi	3	Rutabaga	4
Cauliflower	4	Leek	2	Spinach	3
Celeriac	3	Lettuce	6	Squash	4
Celery	3	Muskmelon	5	Tomato	4
Chard, Swiss	4	Mustard	4	Turnip	4
Chinese cabbage	3	Okra	2	Watermelon	4
Collard	5	Onion	1		
Corn, sweet	2	Parsley	1		

risk to sow parsnip or salsify seeds after they are more than one year old. Two years of sowings can be expected from seed packets of carrot and sweet corn; three years from peas and beans, peppers, radishes, and beets; and four or five years from cabbage, broccoli, Brussels sprouts, cucumbers, melons, and lettuce.

Among flower seeds, the shortest-lived are delphinium, aster, candytuft, and phlox. In general, though, most annual flower seeds are good for one to three years, and most perennial flower seeds for two to four years.

In a frugal mood, I might do a germination test for a definitive measure of whether an old seed packet is worth saving. Counting out 10 to 20 seeds from each packet to be tested, I spread them between two moist paper towels on a plate. Another plate inverted over the first plate seals in moisture and the whole setup then goes where the temperature is warm, around 75 degrees. After one to two weeks, I peel apart the paper towels and count the number of

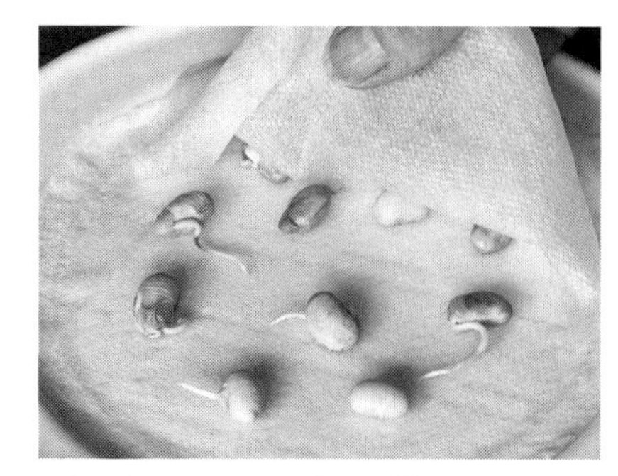

Testing germination of bean seeds.

seeds with little white root "tails". If the percentage is low, the seed packet from which the seeds came gets tossed into the wastebasket or compost pile (I don't give them away!). Or, I might use the seeds and adjust their sowing rate accordingly.

No one knows exactly what happens within a seed to make it lose its viability. Besides lack of germination, old seeds undergo a slight change of color, lose their luster, and show decreased resistance to fungal infections. There is more leakage of substances from dead seeds than from young, fresh seeds, so perhaps aging influences the integrity of the cell membranes. Or, since old seeds are less metabolically active than young seeds, the old seeds leak metabolites that they cannot use.

Electricity temporarily suffices when access to sunlight is lacking.

When God said, "Let there be light," He didn't make quite enough. At least, not enough for raising seedlings indoors in late winter. But way back then, "In the Beginning..." who could have predicted that gardeners in cold winter climates would have wanted to sow tomatoes and marigolds and lettuce indoors to get a jump on the season?

Fortunately, electricity was created, or at least harnessed, over a hundred years ago, and with electricity came artificial light. Early in the 20th century, Cornell scientists embarked on the first experiments in "electro-horticulture," the term they used for growing plants under artificial light, carbon arc lamps initially. Raising seedlings under artificial light is a lot easier and more effective now than it was then.

Good quality, yet inexpensive, lighting for raising top quality seedlings can be had with a combination of ordinary cool or warm white fluorescent tubes, and incandescent bulbs. Why both? Because if you fed sunlight through a prism, and then tested each color separately for its effect on plant growth, you would find that the most effective colors were red and blue. Red and blue light each have their own effects on plants, with, to simplify, red promoting longer stems and larger leaves and blue having pretty much the opposite effect, promoting compact growth. Too much of the former light makes for spindly plants, too much of the latter light makes for stunted plants. Not that green and yellow and orange are without effect, just less so.

Fluorescent lights are rich in blue, with some red. "Cool white" fluorescent bulbs emit very little red light, "warm white" bulbs emit a little more, and "full spectrum" bulbs more still. Incandescent lights are rich in red. Combine fluorescent and incandescent light, and you have a good approximation of sunlight, rich in the most important wavelengths. The combination even looks sunny.

Seedlings could also be raised in the glow of LED lights. Light from an LED spans a very narrow spectrum; if in red, just a narrow band in red, and similarly for blue or any other color. So narrow, in fact, that a different recipe for light seems to be needed for optimum growth of different kinds of plants, or different stages of growth. Performance is generally enhanced with the addition of a small amount of light in the green as well as far-red spectrum (the part of the electromagnetic spectrum just beyond the red that we can see, but shorter wavelength than infra-red). LED "grow lights" are on the market, but research with plant growth under LEDs is in its infancy.

Although the various combinations of fluorescent, incandescent, and LED bulbs offer reasonably good spectral quality, the intensity of these lights does not even hold a candle— pardon the pun—to good Ol' Sol, outdoors. A "foot-candle" is a measure of light intensity, and the sun bathes the Earth with 10,000 foot-candles on a sunny summer day. Indoors, even near the sunniest window, drop that figure to 500 foot-candles or less.

Light intensity drops with the inverse square of distance from a light source, so doubling the distance from a source results in only one-quarter the intensity, tripling the distance results in one-ninth the intensity, etc. Which is to say that seedlings need to be snuggled fairly close to artificial light sources, which aren't all that bright anyway, for best growth. A plant six inches below a standard double fluorescent lamp fixture is bathed in about 600 foot-candles of illumination. Be careful not to put plants too close to incandescent bulbs, however, because these bulbs can generate enough heat to burn a plant. Overall, artificial light works well for seedlings because only a small amount of incandescent light is needed to balance the blues of fluorescent light, and because 600 foot-candles is enough light for them. Double that distance, to about 12 inches, for LEDs.

When I began gardening, I lacked either a greenhouse or sufficient south-facing window space, so I built a phytotron, which is an enclosed space where temperature, light, and other parameters of plant growth can be regulated. I headed down to the local hardware store and purchased two double fluorescent fixtures, with reflectors, and porcelain sockets for incandescent bulbs, along with wire, a plug, a switch, some wood, and assorted fasteners. The reason for two double fluorescent fixtures is because the 600 foot-candles mentioned earlier is the minimum amount of light needed to raise seedlings; a double fixture boosts light levels enough to send seedlings to Winter Plant Paradise. I mounted the porcelain sockets to a 2-by-4 sandwiched between the two fluorescent fixtures, and hung it all on chains for easy lowering and raising. White paint everywhere helped eke maximum light from all sources, as did an occasional dusting off of lights and reflectors.

A good ratio—from the standpoint of a plant—for watts of fluorescent to incandescent light is about 3 to 1. If each fluorescent fixture is 4 feet long, the four 40 watt bulbs offer a total of 160 watts. I balanced that light with three 15 watt incandescent bulbs. A timer turned the lights on for 16 hours each day. Less time would have been needed if the light had been supplemented with natural light through a sunny window, but my phytotron was in the basement.

Exiting my phytotron, after a few weeks of growth, were top notch seedlings. Even so, artificial light is...well...artificial, and not nearly a match for natural light. The sooner plants get out in the sunlight, the better.

In which the pre-plant toughening up of seedlings is shown to be necessary, but with a gentle touch

Indoor or artificial light goes only so far in raising stocky transplants able to withstand the rigors of life out in the garden. Under less than ideal conditions, seedlings stretch out, growing too long and too thin. The combination of a bit too much warmth and a bit too little light causes that stretching.

The easiest way around this problem is to just wait until the weather warms up enough to sow seeds directly outside. There, abundant sunlight, cooler temperatures, and buffeting by wind make for sturdy seedlings. Of course, do this and, in most parts of the country, you'd have to wait until late summer to admire your first zinnia flower or bite into your first tomato.

So we're back indoors. Turning down the heat, pulling window curtains way back, cutting down any trees that block light in a south-facing window—all this helps. But still, light intensity pales in comparison with outdoor light. And the more sun that streams in, the hotter it gets.

Another way to make indoor seedlings sturdier is to merely touch, stroke, or shake them. No need to make this a full time job, because just a few seconds of daily shaking is all that's needed.

Brushing seedlings makes them grow stockier.

I use a brush like the one I use to whisk snow off my truck's windshield, running it lightly over the tops of my seedlings. Sometimes I'll just jostle each tray with a rap of my knuckles. In either case, morning is the best time for the activity.

Although stocky and sturdy growth helps seedlings better survive transplanting and adapt to outdoor conditions, I don't necessarily want my plants to remain dwarfed once planted outside. Fortunately, the dwarfing effect of shaking and touching wears off within days after transplanting.

The fact that seedlings will respond to being touched or shaken is not really all that novel. Buffeting by wind is partly responsible for the diminutive yet tough appearance of a pine tree growing on the windswept edge of a craggy cliff as compared with its svelte counterpart sheltered within the forest.

Shaking and touching plants does not only or always dwarf them. Cucumber or melon plants respond to being caressed by bearing a greater proportion of female flowers. Could this be why cucumber and melon plants growing in the relatively still air of my greenhouse have so many male and so few female flowers? I'll try the snow brush on them also.

Shaking a plant for long periods each day can lead to increased growth, a technique that has been applied in Japanese greenhouses using vibration—even music! I wonder if this means that talking to plants would also affect their growth?

Plants exhibit all sorts of changes, some sought after, some not, as they go through puberty.

Looking at trees that usually drop their leaves by winter, I see that some of them—especially beeches and oaks—wear skirts of foliage all winter long. I say "skirts" because if the trees were human, the leaves would all be at skirt-level. Rather than being lush and green, these leafy raiments are dried and brown or gray, just like their counterparts on the ground.

Trees still clinging to leaves are not out of synch with the environment. Nor does this habit reflect some ecological disaster due to changing climate. The branches cling to their leaves because the branches are "juvenile," and reluctance to drop leaves is one sign of juvenility in plants.

Juvenility in plants is akin to prepuberty in humans: during this period plants grow but are incapable of sexual reproduction, that is, flowering, then setting seed. The duration of juvenility varies from plant to plant. Radish seeds planted in spring will, a few weeks later, send up flower stalks if the roots aren't harvested. Plant an apple seed and ten years, or more, might elapse before the tree first flowers.

Growing conditions influence the amount of time it takes a plant to reach sexual maturity. Wild plants on cold, windswept cliffs grow so slowly that they may still be juvenile after a century. In a greenhouse, with supplemental artificial lighting, apple trees have been coaxed to flower within a couple of years. But no matter how much the scientists fiddled around with growing conditions, no apple plant would flower until its stem was at least 75 to 80 nodes long.

A juvenile plant not only does not flower, but also may have a different form from a mature plant. My father had an English ivy plant that three decades of growth had changed from a creeping vine with lobed leaves, incapable of supporting itself, to a shrub with sturdy stems and rounded leaves—sure signs of maturity in this species. (A horticulture professor of mine once described a gift he received of a "tree" English ivy plant, made by grafting a length of juvenile English ivy onto a robust length of trunk of mature English ivy; the juvenile portion grew vining stems that cascaded down from above the bare mature portion.)

Juvenile shoots also tend to hang onto their leaves, as is the case with those on beech and oak trees. Why, except when young, don't these trees hang on to all, rather than just skirts, of leaves? The reason is because the whole tree isn't juvenile, just the lower branches, which were there when the plant was juvenile. Juvenile portions of a plant always remain so, as do mature portions (whether or not they choose to flower).

You'll never see leafy skirts on any grafted trees, though. Grafting wood is usually taken from mature portions of a stock tree, so a grafted tree is always mature above the graft union. An apple worthy of propagation by grafting is deemed so only because its fruit has been sampled and deemed worthy. It fruited; hence, it is mature.

Does knowing about juvenility make me a better gardener?

Yes, when I propagate plants from cuttings. Juvenile shoots generally root more easily than do mature shoots. Juvenile shoots are those that originate near the base of a plant grown from a seed, or from a cutting made from a juvenile shoot. When I rooted cuttings of my father's English ivy to make more plants, I made those cuttings from the still juvenile shoots growing near the base of his plant. Besides having rounded leaves and shrubby growth habit, mature portions of the plant are also easily discerned from juvenile shoots by bearing (toxic) fruit.

Juvenile growth of paulownia tree.

In most plant species, juvenile shoots grow more vigorously than do mature shoots, and have larger leaves. Juvenile sprouts on paulownia trees often grow 15 feet in one season, with leaves more than a foot across, an effect that is decorative in the right setting. Lopping back all new growth each winter keeps the plant in perpetual youth with a decorative encore each year.

And yes, knowing about juvenility helps when I raise perennial flowers from seed. Perennials usually do not flower until their second season. But by sowing the seeds indoors in March and spurring the plants on with good growing conditions, they can make enough growth to mature and flower their first season.

Mature plants sometimes need just the opposite of the free and luxuriant growth needed by juvenile plants. Fruit trees, although mature (because they are grafted), often are reluctant to flower and fruit. The way to induce a mature plant which is not flowering to do so is with "discipline." Slow down growth by scoring the bark with a knife, by pruning the roots with a spade, or, as described in the "Stems" chapter, by bending upright branches downward.

A recommendation to plant citrus from seed even if fruit is improbable or not worth eating

"What a waste," I was thinking one morning as I spat out a seed from an orange. Not that it was a waste not to eat the seed along with the fruit, but a waste of potential. That seed could grow into a whole orange tree.

Growing an orange tree—or any citrus tree—from a seed is no more difficult than growing a bean plant from seed. As a matter of fact, tangerine seedlings have shared a pot with a houseplant sitting near my rocking chair, evidently "planted" casually as I or someone else ate the fruit while sitting in the chair. It's not unusual to find an overenthusiastic grapefruit seed sprouting while still inside the fruit.

There's only one secret to growing citrus from seed: don't let the seed dry out. Not as critical, but perhaps helpful, would be to soak the seed for a couple of hours before planting to leach out any sprouting inhibitors that might be present. Once the seed has been spat and soaked, it can be planted just like a bean seed, about three-quarters of an inch deep. I've done this in a pot that's been filled with the same potting soil that I would use for houseplants or any other seed. Being subtropical (again like a bean plant), citrus seeds need warmth to sprout. A minimum of 60 degrees Fahrenheit is good enough; 80 degrees would be ideal. Once a seed sprouts, which should not take longer than a few weeks, I move the developing seedling to a sunny window.

Eventually plucking something tasty from any fruit tree grown from seed is likely to be a tenuous proposition. All apple, pear, plum, and peach varieties are clones; trees grown from seed will bear fruits that are different from and frequently worse than—more or less so depending on the particular kind of fruit—fruit from the trees that bore the seeds. That's because seeds, in contrast to fruits, stems, leaves, and roots, represent the commingling and shuffling around of genes of the parent tree with whatever other variety pollinated the flower that preceded the fruit.

However, citrus trees grown from seed can present a more mouth-watering proposition. Citrus has the quirk, known as apomixis, of frequently producing seeds that are not the result of pollination, but that develop from the same kind of cells that make up the rest of the plant. Bingo: a seed that, when grown into a tree, is genetically identical to its mother and, hence, bears identical fruit.

But hold on here: what about Clementines, which lack seeds to interrupt bites into the juicy, sweet segments. Or a seedless Navel tree? Or a tree of any other seedless fruit? You get a new Clementine tree by cloning, and if that sounds too Orwellian then just say "by grafting" or "by cuttings," which are the particular methods of cloning used for most fruit trees.

That answers the question of how you get new Clementine trees, but not the question of how anyone got the very first one. The first Clementine tree may have begun life as a chance or deliberately planted seedling. The genes within this new seedling may just, by chance, have jumbled together into an evolutionary dead end—a tree producing seedless fruits. Another way that Clementine could have begun life was as a lucky, for us, mutation of some branch on a seeded tangerine. Some gardener noticed, tasted, and enjoyed the different fruit on that particular branch, then cloned that branch to make many more new trees, now called Clementines.

(Crunch. I just bit into a seed in this supposedly seedless Clementine. Yes, seeds do occasionally appear, the result of pollen from a different variety of tangerine making its way to a Clementine flower. Clementine is seedless only if grown in isolation. Don't you be tempted to plant any of these seeds, though, because if the seedlings were to bear fruits, they would not, of course, be Clementines.)

A few hurdles still stand in the way 'twixt the soil and the mouth in raising citrus from seed. For one thing, not all the seeds, even in a single citrus fruit, are necessarily apomictic, although sometimes it is possible to identify those produced by pollination by their weaker growth. Or, in the case of the seedling I grew from Flying Dragon hardy citrus, by their characteristic squirmy, twisting stems. Secondly, citrus, like other fruit trees, can take years before they're old enough to bear fruit, especially with less than perfect growing conditions such as in a pot, wintering indoors, in a cold climate. And finally, even after the tree gets old enough to potentially bear fruit, it won't do so except under good growing conditions. Now I'm not saying that it's impossible to provide these conditions to a tree in a pot in cold regions, but you do have to pay attention to providing sufficient food, water, and light.

That said, even a barren citrus tree is worth growing for its glossy, vibrantly green leaves. Growing a citrus tree from a seed is an especially nice long term project for a child. The plants are fast growing and if interest begins to wane, just crush a leaf. The aroma offers a mouth-watering hint of the taste of fruit possibly to come.

Containing some of the ways in which I use a few or many plant cells to conjure up whole new plants

Totipotence is a ten dollar word that refers to the potential ability of any part of a plant, except reproductive cells (egg and sperm) within a flower, to give rise to any other part of a plant, even to a whole new plant. That's because all of a plant's cells (with exceptions, such as with chimeras, in addition to the reproductive cells) house identical genetic information. Depending on the cellular environment and other influences, a cell may become a root, a petal—any part of a plant.

I've made plenty of use of totipotence to multiply a favorite houseplant or shrub, sometimes doing nothing more than dropping some fantail willow stems into a glass of water and watching roots and then new shoots sprout. Such asexual propagation, so-named because it bypasses using seeds (except in the case of apomictic seeds, such as those borne by citrus), results in new plants that are genetically identical to each other and to the mother plant. Totipotence is what lets me start whole new plants from pieces of stem, pieces of root, leaves, or even just a few cells from the growing tip of the mother plant.

Let's start with pieces of stem, like those taken from a coleus plant. Putting the base of a stem into a suitable environment induces roots to form there. Water, although effective with some easy-to-root plants like coleus and willow, is actually not a very good rooting environment. Roots need to breathe, and submerged in water they'll soon be gasping for air. Shaking or changing the water occasionally helps. Roots that do develop in water are structurally different than those that develop in soil and so sometimes have difficulty making the transition from water to soil.

The most effective rooting environment holds moisture and air, and provides support. Nutrients are unnecessary at this point because the stem draws on its nutrient reserves to grow roots and new shoots, and, anyway, lacking roots, the stem would have a hard time drawing up nutrients. Any ordinary potting soil, with a little extra perlite (an inert volcanic rock "popped" at high temperatures) added for better aeration, is suitable. I make up my own rooting mix by combining equal parts peat moss and perlite.

Softwood stem cuttings are lengths of stem, typically a half-foot or so in length, with leaves attached. These cuttings generally root quickly but need care to keep their leaves from drying out before roots develop and draw in more water to support them. I slow water loss from my softwood cuttings by reducing the size or number of leaves and/or reducing water loss from remaining leaves by increasing humidity around them. A covering of clear plastic or glass does this. More high tech is an intermittent mister or, even better, an artificial leaf that actuates a mister when it gets dry, which is ideally about the same time that the real leaves dry out.

Rooted softwood cutting.

Softwood cuttings don't have a lot of energy reserves and need light so that their leaves can feed them. I site my softwood cuttings, which are kept humid beneath a tent or inverted jar, in bright, but indirect, light—in summer, for example, on a bench near the north side of my house. In direct sunlight, they would be "cooked" in their mini-greenhouses.

Hardwood stem cuttings are lengths of stem—usually only a season old—clipped from dormant, leafless (if deciduous) trees, shrubs, and vines. Hardwood cuttings are less susceptible to drying out than softwood cuttings and, if leafless, do not need light, at least until they sprout leaves. Since the presence of leaves can promote rooting, I sometimes take hardwood cuttings in autumn, while stems still retain some leaves, reducing the number of leaves on the cutting to only one or two. Water loss from these leaves is reduced because of the age of the leaves and cool weather with less sunlight.

The procedure for rooting is essentially the same as for softwood cuttings, with two wrinkles. First of all, there's no need to cover deciduous cuttings to increase humidity; they can even be rooted outdoors in any well drained soil. And secondly, cold-hardy woody plants need to experience a certain amount of cool weather before they can grow new shoots. Cuttings taken in autumn and rooted outdoors, as well as cuttings taken at the end of winter, will have experienced this cool period naturally.

With stem cuttings, you start with stems and induce them to sprout roots (and then shoots and leaves). Root cuttings are the opposite: you start with roots and coax them to grow stems (and then leaves and flowers). Not all plants can be propagated by root cuttings but familiar ones that can include oriental poppy, lilac, raspberry, bleeding heart, phlox, and herbaceous peony.

Root pieces that are a couple of inches long and pencil thickness are ideal. The best time to dig them is when plants are dormant, generally in late winter. In the case of oriental poppy and bleeding heart, the best time is in summer, right after the plants finish flowering and go dormant. Small, delicate, root cuttings can be scattered on rooting media, then covered with an inch more of media. Fatter root cuttings are planted vertically in that rooting mix with the ends that were closest to their stems pointing up and just beneath the surface of the mix.

Even leaves can be made to grow roots and shoots. African violet, sansevieria, begonia, and jade plant are among those most amenable to ready multiplication via leaf cuttings. There are a few ways to make leaf cuttings. With rex begonias, just laying a leaf

An African violet leaf makes a whole, new plant.

flat on moist rooting media and scoring across some of the main veins with a razor gives rise to new plants at each cut. With other begonias and with African violets, all that's needed is to poke their leaf stalks, or triangular cut portions of leaves, vertically into a rooting media. A tent of plastic film or an overturned clear, glass or plastic jar maintains the needed high humidity until new plantlets appear.

To avoid the threat of rot, I let leaf cuttings of succulent plants dry out and heal for a couple of days before poking them into moist rooting media, and then let the media thoroughly dry out between waterings. Long, fleshy sansevieria leaves can be chopped into two-inch lengths, then their bottom halves poked into rooting media. The same goes for jade plant leaves, leaving them whole rather than chopping them up. Kalanchoe have the easiest leaves to root, from naturally developing plantlets around the edges of their leaves. These plantlets often drop off and root all by themselves. Succulents need not—and should not—be kept under cover for increased humidity or they're apt to rot rather than root.

I revisit totipotence, using stems again, this time joining them to existing roots.

Joyce Kilmer concluded his poem "Trees" with the lines "Poems are made by fools like me, But only God can make a tree." (And Woody Allen then quipped that the poet wrote this "probably because it's so hard to figure out how to get the bark on.") Well, I can't write a poem, but I can make a tree (with bark attached). It's relatively easy, with grafting.

Totipotence is still at play with grafting, but with grafting there's no need to coax stems to make roots. I just join—that is, graft—a piece of stem, a "scion," atop the stem of an already rooted, growing plant, the "rootstock." The rootstock might be either a very young or a mature plant. All new growth above the graft will be from, and genetically identical to, the scion. I've used grafting to make or make over many of my fruit trees; for instance, to make a new, young pear tree, to add another variety to an existing pear tree, or to decapitate an old pear tree and change it to a better variety. Similarly, a pink dogwood stem could be grafted on a white dogwood tree to create a multicolored flowering dogwood.

The rootstock itself might serve some purpose in addition to merely providing roots. It can influence the form of a plant. Weeping cherry growing on its own roots merely creeps along the ground; grafting it high on an upright trunk of a non-weeping cherry rootstock brings the weeping head high off the ground. A rootstock can influence the eventual size of a plant, as attested to by my 10-year-old Liberty apple tree, topping out at 8 feet high thanks to the 'M.27' rootstock on which it is grafted. Some rootstocks promote quicker bearing, or tolerance to soils that are too wet, too dry, too salty, or nematode infested.

Super-dwarf 'M.27' rootstock dwarfs 10 year old Winesap apple tree.

What's going on that makes it possible to lop the top off a growing plant, lop a stem off another plant, stick the latter atop the former, and have them knit together and grow? Response to lopping when grafting is the same as any other time a stem is cut: A necrotic plate of dried and crushed cells at the cut secretes waxy substances to seal the stem off from water loss and pathogen attack; and the plant begins growing parenchyma cells, which are thin-walled cells plump with water. These parenchyma cells form a callus, which is an undifferentiated mass of cells that eventually breaks through the necrotic plate. As callus of rootstock and scion intermingle, some mechanical strength to the graft is provided, even some flow of water and nutrients between rootstock and scion.

But for long term flow of nutrients and water between rootstock and scion, we need to look near a stem's cambial layer, which is where new cells form. It is from the outer part of this layer that new phloem cells (through which sugars and other metabolic products "phloe" downward in stems) are produced, and, from the inner portion, that new xylem cells (upward conduits for water and minerals) originate. Eventually, if the cambial layers of rootstock and scion are sufficiently close, xylem and phloem cells of the rootstock "find" those of the scion, and functional plumbing is established between root, stems, and leaves of the newly made tree.

No graft will survive without some mechanical strength to keep rootstock and scion together. That strength usually comes from fibers, which are a type of xylem cell. The dwarfing rootstock 'M.9' has very short xylem fibers which provide insufficient mechanical strength to keep a heavily bearing tree from toppling. Mechanical strength of grafts with 'M.9' rootstocks come from the stakes needed to support these tree throughout their lives.

Kinship, contact, and hydration are keys to a successful graft. Rootstock and scion must be closely related—usually, but not always, the same genus and species. Apricot, for instance, can be grafted on peach; they are both in the genus *Prunus*. Cambiums of stock and scion need to be sufficiently close, if not aligned. Drying out of cut surfaces must be avoided, something I do with a coating of a commercial product called Tree-Kote® or with a wrap of Parafilm®. Other products are also available for this purpose.

Given the few prerequisites for a successful graft and grafting's long history—going back to 1,000 BC in China—it's no wonder the many ways that scions have been successfully joined to rootstocks. A monograph by André Thouin in 1821 listed 119 methods. Described below are three easy grafts I have used to make or make over my trees. The ideal time for these grafts is in spring, just as plants are awakening. Scions are best cut ahead of time so that they are still dormant when grafted. I cut scions—pencil-thick, one-year-old stems—in late fall or winter, put them in a plastic bag, wrap the bag in a moist paper towel, then put it all into another plastic bag, well-sealed, in either the refrigerator or a cooler in my garage/barn.

The whip graft is my graft of choice when rootstock and scion are about the same thickness (pencil-thick). I cut off the bottom of the scion with a smooth, sloping cut an inch to an inch and a half long, and make a similar cut at the top of the rootstock. Holding the sloping cuts against each other and aligning just one edge of each if their diameters don't exactly match, I bind rootstock and scion together with a rubber grafting strip (I've also used thick rubber bands, sliced open), then cover cut surfaces to seal in moisture. Parafilm® holds the graft together and seals in moisture. Once a graft is growing strongly, I cut a vertical slit into the binding to prevent it from choking the plant.

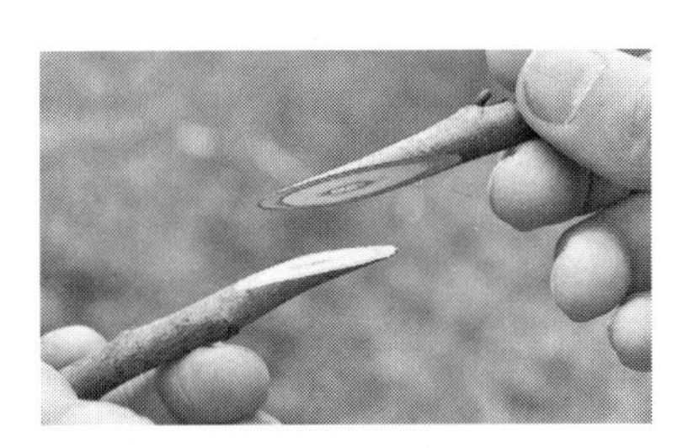

A whip graft, good for joining similarly sized, small scion and rootstock or a stem.

A cleft graft works well when the rootstock is much thicker than the scion, up to about two inches thick.

At the base of each scion, I make two bevel cuts less than halfway through the scion, each two inches long and not exactly on opposite sides, so that, viewed head on and from below, the uncut portion is slightly wedge-shaped. Turning to the rootstock, I lop it off squarely, with a saw, then create a split a couple of inches deep in the middle of the cut surface by hammering a heavy, sharp knife right down into it. After removing the knife, a screwdriver pushed down into the split separates it enough to insert the two prepared scions at each edge of the cleft with their cambiums aligned with the cambium of the rootstock. Pulling out the screwdriver causes the springiness of the rootstock to close the cleft and hold the scions securely in place. I thoroughly coat all cut surfaces to seal in moisture.

A cleft graft, good for rootstocks up to a couple of inches thick.

Even larger rootstocks, up to many inches in diameter, call for the bark graft. This graft comes with an especially good insurance policy. That's because onto each rootstock, depending on its diameter, I can stick 3, 4, 5, or even more scions. Only one scion needs to grow, but the more that are grafted, the greater the chance of success.

I prepare a scion for a bark graft with a bevel cut 2 inches long, at its base, not quite all the way across from one side to the other. On the opposite side of the cut, I nick off a short bevel. Then,

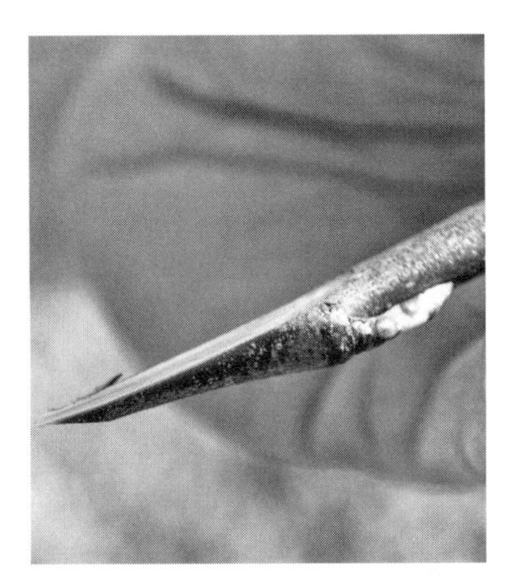

A bark graft, a graft good for large diameter rootstocks.

into the freshly cut stub on the rootstock, I make two vertical slits through the bark, each about 2 inches long and as far apart as the width of the base of the scion. Carefully peeling back the flap of bark welcomes in the long, cut surface of the scion, putting the cambial layers of rootstock and scion in close contact. This is repeated with the other scions, all around the stub. With the peeled back flaps of bark from the rootstock pushed back up against each inserted scion, one or two staples from a staple gun or a tight wrapping with stretchy electrician's tape suffices to hold

Tissue Culture

Tissue culture (micropropagation) is a cloning procedure for multiplying just a few plant cells in test tubes, petri dishes, or other closed, sterile environments before growing them large enough to pot up or plant outdoors. Although generally a laboratory procedure, a home setup can be put together.

To begin tissue culture, cells are teased from a bud or growing tip of a plant, surface sterilized, then placed in the closed container in contact with a liquid or gel medium that supplies the cells with energy, minerals, and a balance of hormones to promote cell division. After a period of undifferentiated division (not yet becoming leaf, root, or any other specific cell type, clumps of cells are transferred aseptically to new petri dishes with a different medium, one that might promote development of roots and/or shoots. Eventually, the small plantlets that develop are transferred to a potting mix in a greenhouse and successively repotted.

Tissue culture is a way to make new plants that, at least initially, are disease-free. And very rapid multiplication is possible in a relatively short time and from just a few cells.

Rooting Hormones

Auxins are plant hormones that, among other roles, play an important role in rooting cuttings. Apply auxins directly to the base of a cutting and there's more chance of the cutting taking root, and rooting might be hurried along. The problem is that natural auxins decompose too quickly to be of practical use once extracted from a plant. Yet auxins can still help cuttings to root—synthetic auxins, such as NAA (α-naphthalene acetic acid) and IBA (indole-3-butyric acid), which are available commercially. IBA is also found naturally in plants, but at very low levels.

Synthetic rooting hormones are available either in liquid or powder form. Concentrations and combinations of auxins vary with manufacturer, with higher auxin concentrations generally used for more difficult-to-root species. Fungicides may be added to prevent rotting of cuttings. Powders are applied by just dipping the base of a cutting into the powder, tapping to shake off excess, then sliding the cutting into a hole dibbled in the rooting mix. Varying amounts of powder might cling to a cutting, leading to inconsistent results. Liquid formulations, besides being more consistent, also are more rapidly absorbed. Merely soak bases of cuttings in the liquid, the time required depending on hormone concentration. Before commercial, synthetic auxins became available, savvy gardeners would help rooting along by soaking cuttings in the water in which the stems of willow, an extremely easy to root plant, had previously been soaked. It worked!

Rooting hormones are not essential in propagation nor do they perform magic. No need to consider using them for propagating willows, chrysanthemums, and other easy to root plants. And no need to waste time trying to root stems from a mature apple or maple tree, or any other plant that just will not root from conventional cuttings.

Rooting hormones also won't make up for poor propagation practices. Close attention is still necessary as to when and what parts of a plant to select for cuttings; the sort of rooting medium used; and moisture, light, and humidity during the rooting and transition period to the real world.

everything in place. Finally, everything needs to be sealed against moisture loss.

Once any of these grafts is completed, I don't just turn my back on it, walk away, and expect success. I always check my grafts the following day and reapply sealant, as needed. The year following a cleft or bark graft, I reduce "takes" to the single, most vigorously growing scion.

Neither monstrous nor scary, but often beautiful—yes, real chimeras may be in our midst.

The chimera of Greek mythology was a scary, fire-breathing creature that was part lion, part goat, and part dragon, and feasted on humans. Although Bellerophon killed that chimera, some still exist today. Perhaps there's one in your backyard, even in your house!

A chimera is a composite creature, a genetic mosaic, and such creatures do exist in the plant world. Don't expect ever to find red apples dangling from marigold stems or gardenia blooms unfolding against backdrops of poinsettia leaves. Plant chimeras are never as genetically diverse as that lion, goat, and dragon combo. Nor are they as physically diverse, a plant usually broadcasting that it is a chimera only with splotches or lines of color different from the surrounding color of the leaves, flowers, or fruits.

A chimera might originate by design, but more usually by chance. To picture the beginnings of such a creature requires a step back to thinking how any plant grows.

All plants elongate by division of cells at the tips of their stems. Zoom in to one of those stem tips, down to the cellular level, and you'll see that it has two or three well-defined layers which, as they divide, give rise to distinctive parts of the plant. For instance, the outermost layer of the tip becomes, logically enough, the outermost layer of a leaf. In most plants all the cells in their tips are genetically identical to each other and to those in the rest of the plant, with the exception of the pollen and egg cells.

Now just suppose that a portion of that stem tip—even just a single cell—is genetically a bit different from the others. Perhaps that cell and its offspring were colorless. Then whatever parts of stems, leaves, flowers, or fruits derive from that particular cell would also be colorless.

That oddball cell or cells could be the result of a natural mutation. Or, a stem tip with more than one kind of cell could be made by tissue culture. Historically, gardeners have occasionally created chimeras when grafting if, by chance, a new growing point arose that incorporated dividing cells from the two parts of the graft.

Some of these so-called graft hybrids aspire to be like the chimera of mythology, not in fierceness but in creating a creature representing more than one species or even genus. The camellia 'Daisy Eagleson' is a graft hybrid of two different camellia species. Graft hybrids have also resulted from grafting laburnum and broom plants together, which are in different genera although the same family. The resulting plant's branches are usually draped in yellow flowers characteristic of laburnums, but occasional branches are covered with purple blooms of broom.

The plant chimeras that we gardeners are most familiar with are those that are visually obvious and look pretty—how else would we so easily identify them, and why else would we be so ready to propagate them? Thus we have the vinca varieties 'Elegantissima' and 'Oxoniensis', the former with white margins bordering a dark leaf and the latter with dark margins bordering a pale green leaf. Another plant chimera is the sansevieria variety 'Hahnii Solid Gold'. Or a citrus fruit with alternating orange and yellow skin segments. Chimeras are relatively common among geraniums.

Don't assume that any plant with streaks or splotches of color is a chimera. Wild colorations are sometimes the effect of a virus, or of nutritional or environmental problems. And sometimes—in lungwort or zebra plant, for example—certain cells naturally grow differently or take on a different color in certain areas of the plant even though the whole plant remains genetically homogeneous. Perhaps such plants just want to look like fierce beasts.

Knowing that a bulb is, essentially, a stem lets me multiply them with the same "pinch" that makes stems branch.

Knowing what a bulb really is makes it easier to understand how they can multiply so prolifically that their flowering suffers, and how to get them to multiply for our benefit. But first, not all "bulbs" are bulbs. They're all underground structures in which a plant stores food to carry it through a dormant period. A gladiolus, a crocus, or a crocosmia, for example, might be called a bulb, but, in fact, is another structure—a corm, which is a swollen, underground stem with a compressed area of growing points, which might include flower buds.

True bulbs are a different stripe of underground storage organ. Dig up a tulip or daffodil—both of which are true bulbs—and slice it through the middle from the tip to the base. What you see is a series of fleshy scales, which are modified leaves, attached at their bases to a basal plate, off of which grow roots—just like an onion (but, in the case of daffodil, poisonous!). The scales store food for the bulb while it is dormant.

That bottom plate is the actual stem of the plant, a stem that has been telescoped down so that it is only about a half-inch long. Those fleshy scales are what store food for the bulb, and, depending on the kind of bulb, are nothing more than modified leaves or the thickened bases of leaves.

Look at where the leaves meet the stems of a tomato vine, a maple tree—any plant, in fact—and you will notice that a bud develops just above that meeting point. Buds likewise develop in a bulb where each fleshy scale meets the stem, which is that bottom plate. On a tomato vine, buds can grow to become shoots; on a bulb, buds can grow to become small bulbs, called bulblets. In time, a bulblet graduates to become a full-fledged bulb that is large enough to flower.

Let's see just how the big three flowering bulbs—tulips, daffodils, and hyacinths—grow. With tulips, the main bulb, the one that flowers, disintegrates right after flowering and is replaced by a cluster of new bulblets or bulbs. The mother bulb of a daffodil continues to grow after flowering, at the same time producing offsets of bulbs or bulblets which may remain attached to the mother bulb. Hyacinths grow the same way as daffodils, except for being much more reluctant to make babies.

All those bulblets and bulbs can, in time, crowd each other for light, food, and water, so flowering suffers. When my daffodils, tulips, or other spring-flowering bulbs seem to be making too many leaves and too few flowers, I know it's time to dig them up and separate all the bulbs and bulblets. The best time for this operation is just as the foliage is dying back in spring or early summer because after that plants are out of sight until the following spring. Replanting can be done immediately or delayed until autumn.

Knowing something about bulb structure also gives me inside information on how to multiply my holdings, perhaps to make enough plants for a whole field or a giant flower bed. Merely digging up and planting offsets is one way of doing this. Also, knowing that the bottom plate is a stem and that the offsets come from side buds lets me coax those side buds to grow just as can be done with any other stem.

Pinching out the growing tip of any stem makes side buds grow. One way you get rid of, or "pinch out," the growing tip of a bulb is to turn the bulb upside down, then scoop out the center

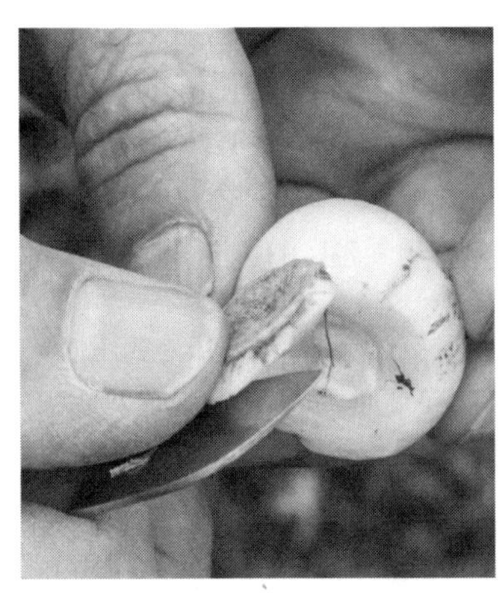

Scooping hyacinth bulb.

Scoring hyacinth bulb.

of the bottom plate with a spoon or knife. Another way is to score the bottom plate with a knife into six pie shaped wedges, making each score cut deep enough to hit that growing point. After scoring or scooping, set the bulb in a warm place in dry sand or soil for planting outdoors in a nursery bed in autumn.

After a year, as many as 60 bulblets might be sprouting from the base of each scooped or scored bulb. Each bulblet does take four or five years to reach flowering size, but no matter. To quote an old Chinese saying: "The longest journey begins with the first step."

SOIL

In which we watch the progress of water traveling through soil, with methods to, at the same time, speed it up and slow it down

My friend Mike gets a chuckle from the way I so consistently diagnose his houseplant's troubles. "Too much water," I always say, at first. And then, if Mike protests that he hardly ever remembers to water a sickly plant in question, I merely respond, "Too little water." (Again, a chuckle.) Water, too much or too little, is often the root (pun intended) of a plant's problems whether in a pot or out in the garden.

Let's dig down into the soil to look more closely at what's going on there, waterwise. Ground up rocks, or minerals, make up about half the volume of most soils. A few, but important, percent of the volume is organic matter (stuff that is or was living), and the rest is pore space which can fill with water and/or air. The water is needed to move nutrients into and through the plant, to cool it, and to keep cells plumped up. Less obviously, air is needed because, like you and me, roots need to "breathe," that is, take in oxygen and release carbon dioxide. Problems arise when too much or too little water occupies the pore space. Either extreme presents the same symptoms because plants have no access to water if either it's not present or if the plant's roots aren't functioning; hence my diagnoses for Mike's houseplant problems.

Water drainage is so important that I give it top priority when assessing a potential planting site. Wetland plants such as purple loosestrife, buttercup, cattails, or cardinal flower tell of drainage problems. More quantitatively, I can measure drainage in summer by removing the top and the bottom from a straight-sided coffee can and, after digging a hole a few inches deep, sinking the can into the hole with its bottom edge pressed firmly into the soil at the bottom of the hole. I fill the can with water, let it drain, then fill it again, and measure how long it takes for the water level to drop. Anything slower than 1 inch per hour indicates a drainage problem.

Measuring water drainage of soil.

The first of two causes for drainage problems is a permanent, high water table. One workaround is to simply lower the water table, draining the water away to lower ground via open ditches or buried, perforated plastic pipes. When I worked as a soil conservationist for the US Department of Agriculture Soil Conservation Service (now called the Natural Resource Conservation Service), we surveyed farm ditches five feet deep and sometimes six feet across. No need to gouge out such drama in the landscape for a garden: a shovel's width or two is sufficient on a garden scale. The deeper the ditch, the greater the resulting depth of well-aerated soil, so a ditch should be at least 18 inches deep. A gradual slope along the length of any ditch, about a half foot per hundred feet, keeps the water within flowing downhill.

When I worked as a soil conservationist, I also helped with another work around for a high water table, designing and supervising the digging (with large trenching machines) of deep trenches into which was laid "drainage tile," an archaic name for perforated, four-inch-diameter plastic pipe. (In decades past,

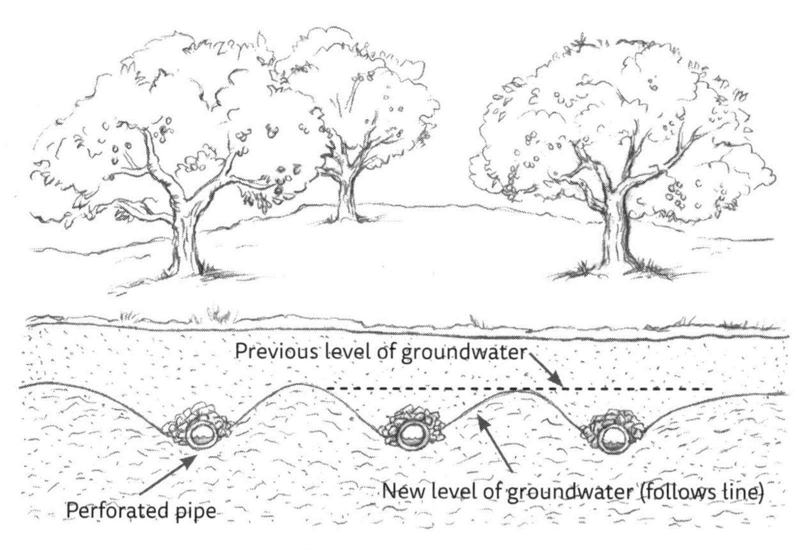

Soil drainage with "tile."

drainage tile really was tile, consisting of short lengths of clay pipe laid end to end with a small gap for water to enter between adjacent lengths.) After being set in a ditch, the pipe was covered with some fabric or paper to keep dirt out of the holes, topped with a layer of gravel and then backfilled with soil. A screen at the outlet end of the pipe kept out curious animals. "Drainage tile" does away with having to maneuver over and maintain ditches throughout a field or trenches in a garden.

One length of perforated pipe could not drain a large field, so connecting networks of perforated pipe are needed. A single line or two might be all that's needed on a garden scale, with more lines and closer spacing needed to draw water from slowly percolating clay soils. As with a ditch, the tile needs to have an outlet in lower ground, to which it needs a downward slope.

Instead of lowering the water table, I sometimes raise my plants' roots above the water table to rescue them from seasonal drowning. Either raised beds or wide mounds work well, the former more useful for vegetables and flowers, the latter more so for trees and shrubs.

The second cause for poor drainage relates to a soil's texture and structure. Texture is the particle size distribution of the

Blueberries on raised mounds for poorly drained soil.

minerals, and structure is how those particles are aggregated into larger units. Soil minerals are grouped into three size categories: clay particles are the smallest, by definition less than 0.002 mm; sand particles are the largest, larger than 0.1 mm; in between these extremes are silt particles. The proportion of sand, silt, and clay in a soil is referred to as a soil's "texture."

Texture and structure influence water drainage because smaller particles have smaller pore spaces between them. The smallest pores are capillary-size, so can draw in water against the pull of gravity by capillary action, that is, by cohesion of water molecules to each other and adhesion of water molecules to the surfaces of the soil particles. The largest soil pores are too large for capillary action to hold them full of water, so gravity can pull water down and out, in so doing drawing air into the pores.

Few soils are made up of only one size of particles. Depending on the contribution of each size fraction, a soil may be assigned

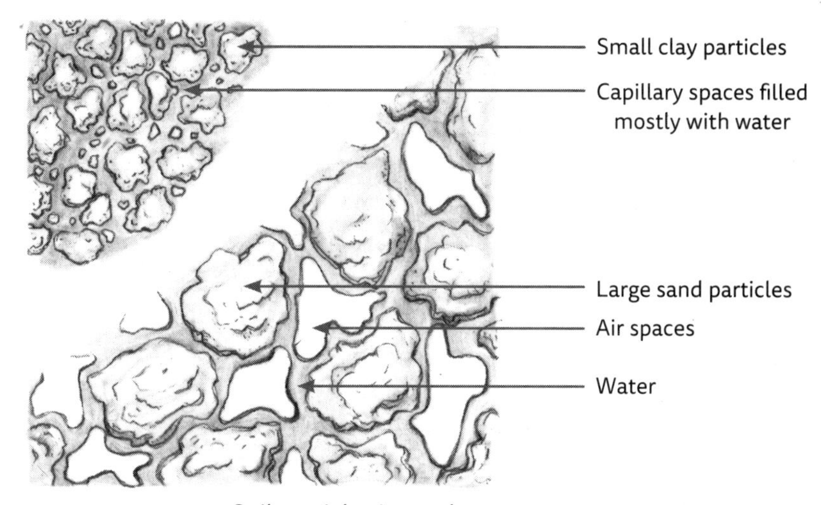

Soil particle size and water.

a name—a "textural class" such a sand, sandy clay, or silty clay, etc. The "textural triangle," each side scaled according to the percentages of sand, silt, and clay, schematically shows the makeup of each of the twelve soil textural classes. Soils with about equal functional contributions from sand, silt, and clay are classed as

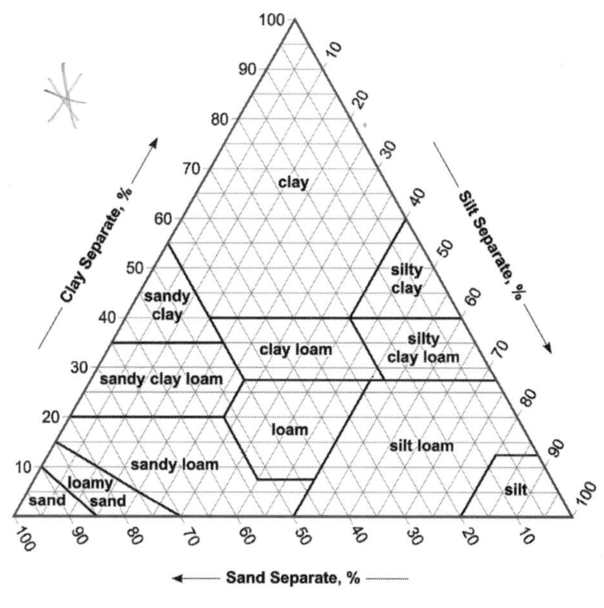

Soil textural triangle.

"loams." Even within the category loam, though, those that show more influence from one or more of the size fractions might, according to the contribution of the fraction, be assigned to such classes as clay loam, sandy loam, and sandy clay loam.

What happens when water is added to a soil having a range of pore sizes? Following heavy rain or watering, all pores fill with water and the soil becomes "saturated," not a happy state for roots but, hopefully for them, a temporary one. Once rain lets up, gravity drains water from large pores to bring the soil to "field capacity." In this root-happy state, roots breathe freely the air held in large pores, and have easy access to both thick films of water on soil particles and pockets of water in smaller pores, both held only weakly by capillary forces.

As roots drink up water, the remaining water becomes harder and harder to access because roots must pull against capillary forces that become stronger and stronger. Water is held in soils in small pores and as a film on larger particles. Water is held more tightly to the particles as the water film grows thinner. Eventually, in the absence of rain or watering, the soil dries down to a point where water, though present, is held too tightly for plants to draw in. "Wilting point" is reached, after which, unless additional water percolates into the soil, "permanent wilting point" is reached, from which a plant cannot recover. "Plant available water" is the difference between field capacity and permanent wilting point.

The ideal soil, it would then seem, would be a loam, its range in particle and, hence, pore sizes able to hold an agreeable amount of air and water. True. But no gardener needs to "grin and bear it" with whatever soil pore spaces are found underfoot. While it's not usually practical to change a soil's texture (adding sand to some kinds of clays makes concrete!) soil particles can be reorganized for better porosity. "Aggregation" is the clumping of small particles into larger units; larger units have commensurately larger pores between them. So a fine clay that might otherwise be too goopy from clinging tightly to so much capillary water becomes

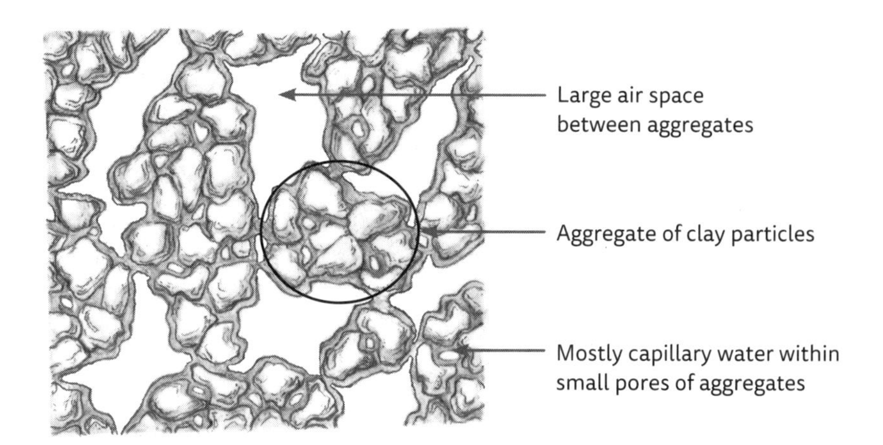

Soil aggregation.

well aerated when aggregated. The result is the best of both worlds: good aeration from the large pores between aggregates, and a good drink of available water from some of the smaller pores (but not the very smallest pores) within the aggregates.

If a soil is left alone, root growth, freezing and thawing, wetting and drying, and the activity of micro- and macro-organisms, will, over time, aggregate it. But no one wants to wait that long. What's more, many garden and farm activities, such as leaving the soil surface bare, digging or tilling the soil, even walking or driving (including tractors) on the ground all have the potential to de-aggregate a soil.

Especially in the western US, sodium in soils prevents their aggregation, as relatively large sodium ions insert themselves between clay particles to keep them dispersed. Gypsum (calcium sulfate, twenty pounds per hundred square feet) can alleviate the condition, the relatively small calcium ion, along with plenty of water, flushing away the sodium ions to let the clay particles aggregate.

We're not finished with soil aggregation and will return to it shortly, along with dealing with sandy soils, which tend to be too well-aerated and too dry.

Hand Texturing Soil

Grubbing around in the garden, planting, tilling, and weeding, can't help but give me some feel for my soil's texture. I cursed the stickiness of the clay soil of my first garden, in Wisconsin, and how quickly the sandy soil in my second garden, in Delaware, dried out. These two soils were extremes. "Hand texturing" is a quick and easy way to even better know a soil's textures.

I hand texture my soil by first gathering a tablespoon or so of it in my palm. Adding water slowly until it has the consistency of modeling clay, I form it into a ball. If it can't hold together as a ball, it's a "sand." Otherwise, I start kneading the ball and then squeeze it out between my thumb and forefinger into a flat ribbon that eventually collapses under its own weight. The length and feel of the ribbon tells the texture. If no ribbon develops, the soil is a "loamy sand."

Hand texturing a soil.

More clay makes the ribbon develop longer. If the ribbon breaks after less than an inch, the soil is a "loam" or "silt," between one and two inches signifies a "clay loam," and longer than two inches signifies "clay."

Sand feels gritty, silt feels silky smooth, and clay feels sticky, so with more attention to the feel of the soil, I can hone the classification of what's in my hand into one of the twelve soil textural classes. For example, if a ribbon feels gritty and breaks when it's between one and two inches long, the soil is a "sandy clay loam." My present garden is a "clay loam."

Sedimentation Test for Soil Texture

A sedimentation test, which capitalizes on the fact that large, heavy particles sink more quickly in water, provides a more quantitative measure of soil texture than does hand texturing. All that's needed is a quart Mason jar, water, Calgon® detergent (or baking soda), a ruler, and a timer. The Calgon® or baking soda de-aggregates the particles; otherwise, clumps of clay particles might register as sand or silt.

Add about ½ cup of soil, dried out for a day or two and sifted through a screen or a colander to remove sticks, stones, and other debris, to the jar, along with 3½ cups of water and a few tablespoons of Calgon® or baking soda. Capping the jar, shake it very well for 5 minutes, then let it sit undisturbed for 2 days, during which time all soil particles will settle. Measuring the depth of settled soil with a ruler gives "total soil depth."

Shake the jar again, for 5 minutes, this time letting it stand for 40 seconds to let all sand particles settle out. The percent sand is the sand depth divided by the total soil depth, times 100.

Let the jar stand for 30 minutes more, without shaking, which is long enough to let all silt particles settle out. The percent of silt is the sand plus silt depth, minus the sand depth, divided by the total soil depth, times 100.

The percent of clay is the total soil depth minus the sand plus silt depth, divided by the total soil depth, times 100.

Knowing the percentages of sand, silt, and clay, you can trace the measured percentage of each soil particle down the appropriate line from its side of the textural triangle to assign the soil to a textural class.

A common sense recommendation that turns out not to make sense

Things get more complicated when water must move down through a soil in which there is a change in porosity at some depth—a not uncommon occurrence in the field, in a flowerpot, or, deliberately, in a tree planting hole. A layer of clay soil might, for instance, overlay one of sand. Two forces are at work in moving water within soil: gravitational forces and capillary forces. Capillary forces are the stronger of the two forces and, as previously mentioned, are greater the closer water molecules are to soil particles. So the drier the soil, the more prominent the capillary forces and the slower the movement of water.

Picture a porous sand overlaying a not-so-porous layer of clay. Gravity readily pulls water down through the large pores of the sand. That downward flow screeches to a near standstill when the water encounters the clay layer. Although capillary forces draw water down and laterally into the clay layer, that first water that enters the clay layer is held tightly against the pull of gravity. Because the water flows much more slowly through the small pores of the clay, a wetting front builds up in the sand, backing up at the base of the sandy layer. The result is what's called a perched water table, the height of which depends on the relative porosities of the two layers.

Next, picture the opposite: a clay overlying a sand. Common sense would presume that the sand layer would speed up water drainage through the sluggish clay. Not so! As water is added to the clay layer, capillary forces pull on the water in all directions wherever the water reaches drier soil, the wetting front. When that wetting front encounters the coarser, sand layer, downward flow, as in the previous example, again screeches to a near standstill. Sandy soils, with large pores, exert little capillary suction to pull in the water from above. Not until the clay layer is sufficiently saturated will it have any water held loosely enough to allow gravity to pull it further downward. And even then, flow down through the sandy layer is slow because of the slow entry of water into the clay layer from above. Again, the result is a perched water table whose depth, again, depends on the relative porosities of the two layers. The smaller the pores in the clay layers, i.e., the more clayey the upper layer of soil, the deeper the perched water table.

A misunderstanding of the dynamics of water movement has led to some misguided recommendations for gardeners. One recommendation is to put coarse material, such as pieces of wood, or gravel, into the bottom of a tree planting hole "to improve drainage." The resulting discontinuity in porosity creates above it a perched water table, i.e., even worse drainage! The effect is the same for following the recommendation to put a coarse layer of gravel in the bottom of a flowerpot, again "to improve drainage." The resulting perched water table decreases the depth of well-aerated soil the roots have to explore.

A flowerpot or seedling flat generates an unavoidable perched water table because a gross discontinuity in porosity cannot help but exist near or at the base of the container. If it's not between the potting soil and the air beneath the pot, it's between the potting soil and the solid bottom of the pot, or between the potting soil and the flat of the saucer beneath the pot.

Perched water table in pot with more porous lower layer.

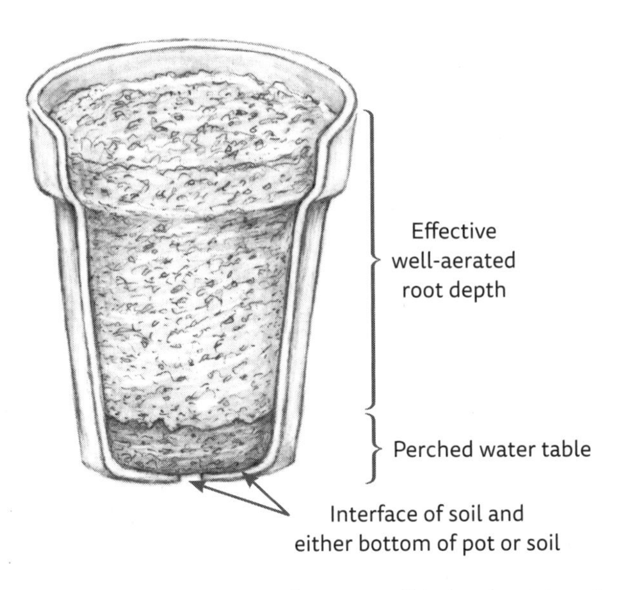

Perched water table in flowerpot filled only with soil.

The unavoidable perched water table at the bottom of any flowerpot is what makes straight up garden soil, even good garden soil, unsuitable for potted plants: capillary attraction of relatively dense garden soil draws water up from a perched water table, further limiting the zone of good aeration. I add "aggregate," such as perlite or vermiculite, to my potting mixes. These very porous materials reduce the capillary forces in the potting mix to keep water from the perched water table from climbing too high in the potting mix. A piece of window screen over the drainage holes at the bottom of flowerpots is all that's needed at the bottom of the pot, just to keep the potting mix from running out the holes. A porous potting soil and a screen (to keep soil from flowing out the drainage hole) put the perched water table as low as possible in the pot, and its depth to a minimum.

(A "good garden soil" is good for the garden even if it's not good in a flowerpot, as long as any perched water table from a discontinuity in soil porosity lies deep enough in the ground to give roots a good depth of aerated root run.)

Contains a description and an opinion of hydroponics

Pushing my cart down the produce aisle in the grocery store, I come upon tomatoes, and on each of their red skins is a small tag advertising(?) "hydroponically grown." Hydroponic plants do not grow in soil. The roots, instead, find their home in some inert medium like vermiculite, sand, or gravel, which serves no other purpose than to support the plant. Or roots might be continuously bathed in an aerated nutrient solution, in which case some sort of support for the plant must be devised.

Humans have long been drawn to high tech, and the "bells and whistles" (pumps, bubblers, etc.) of hydroponics satisfy that need. But why would the farmer who grew the hydroponic tomatoes I examined in the grocery store last week want to grow plants hydroponically? What could be the attraction, given that real soil is a renewable, natural resource that has done a good job of supporting plant life for eons?

One reason for commercial hydroponics is to avoid pests, which are a threat especially to greenhouse growers. Theoretically, if a greenhouse grower started with "clean" plants growing in an inert, sterile medium like gravel, then fed the plants an inert, sterile nutrient solution, there would be no pests to plague plants—except when a disease spore happens to waft into a hydroponic system. Nutrient solution sloshing from one plant to the next is very effective at spreading pests once one plant is infected. And in the absence of beneficial microorganisms, which are present in soil under natural conditions, pests that gain a foothold can multiply unchecked.

The claim is made that hydroponic plants yield more than plants growing in real soil. Present claims are nowhere near as extravagant as when hydroponics was first developed. That was in 1929, and newspapers of the day hailed hydroponics as the greatest invention of the century, predicting that farmland soon would become a relic of the past. Such claims now are tempered.

I contend that if the same care was lavished on conventional plants as on hydroponic plants, discrepancies would vanish. Also, yield is not the only consideration in growing plants. Nutritional quality and plant health also are important, and not always directly related to yield.

Hydroponics does make it feasible to grow plants where there is no soil. During World War II, fresh vegetables for American G.I.'s were raised in this manner on non-arable islands in the Pacific.

But real soil, wherever it exists, can nourish plants naturally. If I scoop out a mere teaspoon of soil, that teaspoon is home to 10,000 to 50,000 species and more microorganisms than people on earth. Several yards of fungal thread course through that teaspoon. A handful of soil has more biodiversity than all the animals in the Amazon basin. Those burgeoning populations are not there for naught. Some microorganisms gobble up plant pests; others chew up dead leaves, stems and roots of plants, and dead animals, in so doing recycling nutrients for use by living plants. Still others use their long, thin, thread-like bodies to gather nutrients for plants from the far-reaches of the soil. All these organisms also produce humates, hormones, antibiotics, and chelators, which are not nutrients, but do affect plant growth and health.

Nothing like this goes on beneath a hydroponically grown plant. Hydroponic plants get only the 15 or so nutrients deemed essential for their growth, and all nutrition comes from chemicals dissolved in water or, more recently, from the water of aquaponic tanks used to raise fish. The hydroponic farmer has to go out and buy these nutrients, along with pumps and tanks and timers, and electricity to run the pumps and timers. Consequently, hydroponics can be costly to install and to run.

Hydroponics is not very forgiving. Divorced from the natural soil environment, plants get what you decide to give them; my faith in science has limits, and doesn't extend to believing that present knowledge is at the point where a batch of chemicals can be concocted that can optimally nourish plants or people. Contemporary hydroponicists bridge the gap to "organic" agriculture by using, for nutrient solution, nutrients derived from organic sources. But what goes on below "ground" for a hydroponically grown plant is still very different from what goes on in real soil.

Hydroponics does have the advantage—or is it a disadvantage?—of being amenable to automation. Once the hydroponicist has a system perfected for a particular crop, pretty much all that's needed is for someone to stare at some dials to make sure everything keeps operating smoothly. But that can present a problem—boredom. Much more satisfying, to me, is to spread my witch's brew of compost on the ground to grow healthy plants.

In which I pay homage to humus, even though it may be a misnomer

I promised to return to aggregation and soil water. Now I will, with this homage to humus (HYOU-muss, not to be confused with hummus, HUM-mess, the delectable Middle Eastern spread or dip of chickpeas and tahini).

No one can say exactly what humus is because it's a witch's brew of myriad carbon-containing compounds that result from the decomposition of dead plants and animals. "Yuk!" Not so. Compost, leaf mold, or the spongy, dark layer of earth that's exposed when leaves lying on the forest floor are pushed aside are

Midwest prairie soil, once perennial grasses, now corn.

mostly organic compounds. Humus-rich soils call to mind our Midwestern plains, the Argentine pampas, the Russian steppes. Humus is what has made such soils the breadbaskets of the world.

Both the chemistry and the feel of humus make it such great stuff. For instance, humus is covered with negative charges, and these charges keep positively charged nutrients, such as potassium and calcium, from washing out of the soil.

A soil rich in humus is also rich in microbial glues. At first, glue of any kind might seem like a bad thing for soil, but these glues are what join small clay particles into the larger aggregates that have commensurately large air spaces between them. Lo and behold, that tight clay soil is then breathing as easily as a well-aerated sand. Regularly feeding my soil with organic materials, which become humus, is part of my quiver of methods for improving soil structure.

Humus also brings other benefits, such as buffering acidity, which frees me from having to be finicky about getting my soil pH exactly right. And humus binds with certain nutrients—iron, for example—to facilitate their absorption by plants. Physically, the sponginess of humus makes soils fluffier even as it absorbs water—just what plants like. Humus bleeds plant nutrients into the soil as it decomposes, a whole slew of nutrients rather than just the major ones lying within a bag of 10-10-10 or most other chemical fertilizers.

* * *

I can't write "humus" even one more time without a caveat. Humus has always been considered to be the stable portion of soil organic matter, the portion pretty much resistant to further degradation. That view has recently been challenged with what has been called "humus" now branded as a poorly defined artifact of experimental science—basically, what's left that won't dissolve in a relatively strong alkaline solution. This residue does not exist in nature; only after laboratory treatment.

The new soil continuum model, positing a continuum of "progressively decomposing organic compounds," has both agricultural and environmental implications. Rather than trying to build up stable reserves of humus in the soil, we need to pay attention and manage flows of organic matter in the soil. Henceforth, I will abandon the word "humus" and refer, instead to "soil organic matter," representing the continuum in the soil of many different organic compounds in various stages of decay. I will continue, now, with my Homage to Soil Organic Matter.

* * *

Soil organic matter is one of those few things in life that you can't get too much of. Although it's naturally present in all soils at levels typically from one to five percent, good gardening mandates a conscientious preservation or, even better, beefing up of those levels. This is because many garden activities hasten soil organic matter decomposition. (Besides the effect on soil, decomposition of soil organic matter sends the greenhouse gas carbon dioxide wafting into the atmosphere; conversely, building up levels of soil organic matter takes CO_2 out of the atmosphere.)

Not that soil organic matter decomposition is a bad thing; many of its benefits accrue during its decomposition. But when soil organic matter loss outstrips its rate of accumulation, it's like taking money out of a bank faster than you put it in.

Tilling a soil is one garden activity that promotes loss of soil organic matter. That stirring of the soil charges it with air, causing microbial populations to soar, and these hungry microorganisms then gobble up soil organic matter. Following an initial burst of nutrients, the soil is left poorer. Tilling organic matter into a soil adds organic matter while, at the same time, promotes decomposition of existing organic matter. The net result depends on variables such as the amount of tillage, the amount and kind of organic matter added, and the soil texture. Not that tilling is necessary; many farms these days practice no-till or minimal till, and I haven't tilled my garden for over 30 years.

Besides air and carbon compounds, the other major limitation for microbial growth in soils is nitrogen (not nitrogen as found as a gas in air, but combined with other elements). Concentrated nitrogen fertilizers have a similar effect to tilling in speeding up the "burning" up of organic matter. With neither nitrogen nor oxygen limiting their growth, microorganisms can make quick work of the carbon in soil organic matter.

I regularly feed my soil organic materials, either hauling them in or growing them in place. Wood shavings, autumn leaves, straw, pine needles, sawdust, and wood chips are readily available, often as waste products. The sandwich board on the street in front of my house that reads "WOOD CHIPS WANTED" gives local arborists passing by the opportunity to unload their truckloads of "waste" wood chips in my side yard. Some gets spread as mulch, some finds its way into my compost bins. Similarly for hay that I scythe from my meadow, bags of leaves that neighbors rake from their yards and then call me to haul away, and horse manure from a nearby stable. Destined only for my compost are vegetable trimmings from the kitchen, old plants and weeds from the vegetable garden, even my old cotton jeans and underwear. Anything that is or was living, or was made from something that was living, gets put back on the land.

Occasionally I'll grow organic matter in place, planting a cover crop in a vegetable bed no longer slated for planting near the end

Oat grain cover crop in vegetable garden.

of the growing season. Cover crops are plants grown not for harvest, but specifically to improve the soil. In theory, at least, the benefits of organic matter in the soil can be had without any hauling, by merely growing the organic matter in place, which is what cover crops do. (Whether or not benefits accrue depends on what's grown and when and how it's managed; it might be tilled into the soil or the above-ground portion might be left on the ground as mulch following a natural death or death by timely mowing.) I've mostly sown oats, or a mix of oats and peas. These plants enjoy cool weather so even late summer sowings create a verdant carpet by December. Oats and peas are snuffed out by February's cold where I live so don't need to be tilled to kill them and make way for planting the following spring. Another option is to set aside a portion of land or a portion of the main growing season for cover crops. Grassy plants, such as oats, rye, sorghum, and wheat, are the best for increasing soil organic matter.

Of course, I've only scratched the surface of the ground (ha, ha) in this homage to soil organic matter. Once reverence is gained for this material, it pays to explore ways to preserve and augment it. Soil organic matter is what put the "organic" in organic gardening, but a soil rich in organic matter is the earmark of any good gardener.

Wherein I check my ground's acidity and then tweak it, as needed

Horrendous sounding afflictions like "verticillium" and "fusarium" roll so easily off the tongue. Which is why, perhaps, these or other diseases are blamed for the sickly yellowing of a pin oak's or geranium's leaves. Consider, instead, that soil acidity might just be out of whack. Every plant has its preferred range of soil acidity and when acidity falls out of that range, a host of ills follow.

Acidity, whether in the soil or in a beaker in a chemistry lab, is measured in pH units, a scale running from 0 to 14. What pH is measuring is the concentration, in solution, of hydrogen ions (H+). Acidity of soil comes from hydrogen ions attached to

Cover Crops, Other Benefits

I blanket my vegetable garden with more than enough compost each year, so increasing soil organic matter would hardly be sufficient reason for me to sow a cover crop. Cover crops bring other benefits. Densely sown oats create a commensurately dense, verdant carpet that's effective in shading out weeds. That verdant carpet also protects the surface of the ground. From what? Rain battering bare soil can seal the surface, preventing percolation of water and, if there's a slope, washing soil particles downhill. The verdant carpet also insulates the ground against wide swings in temperature, all to the liking of earthworms, fungi, bacteria, and other soil life.

Dropping further below the surface, roots of cover crop plants, as they ply through the ground, break the soil up for better movement of air and water, a legacy also of roots that have died and decayed. Further down in the ground, soluble nutrients might be washing away beyond the reach of roots, perhaps contaminating groundwater—unless latched onto and brought back up to shallower ground by the growing roots of cover crop plants.

The most obvious benefit I get from cover crops is their lush greenery that persists all through autumn, particularly welcome when most other colors are fading from the surrounding landscape.

mineral particles and to soil organic matter. Actual acidity depends on the amount of organic materials in the soil, as well as the kinds and amounts of minerals and what we add to the soil.

Because hydrogen ion concentration varies over such a wide range, it's measured on a logarithmic scale, which means that a one unit change of pH corresponds to a 10-fold change in hydrogen ion concentration. Okay, it does get a little more convoluted:

Nitrogen Starvation

I see that someone just "raised their hand" to point out that adding fresh, low nitrogen, organic materials to the soil results in nitrogen starvation of the plants. This science-y, oft-repeated (and printed) myth needs debunking.

As stated above, with sufficient aeration, soil bacteria and fungi have greatest hunger for two nutrients: nitrogen and carbon. Just as adding concentrated nitrogen fertilizer to a soil speeds decomposition of soil organic matter, adding a high carbon material such as wood chips and leaves increases demand for nitrogen. Microorganisms, then, grab at any other nitrogen in the ground to eat along with their fresh chips or leaves.

Because microorganisms are better at garnering soil nitrogen than plants, plants get starved for nitrogen. Only temporarily, though, until some of the digested carbon is given off as carbon dioxide and what's left are the higher nitrogen dead remains and excreta of bacteria and fungi.

However, when fresh chips or leaves are used as mulch, decomposition proceeds very slowly at the interface of soil and mulch. So slowly that nitrogen is re-released into the ground at about the rate it's being tied up. Digging chips or leaves into the soil will definitely cause a temporary tie-up of nitrogen; mulching with these materials will not...yet another reason to avoid digging or tillage.

to keep the numbers more conveniently positive, the sign of the numbers is changed. So a hydrogen ion concentration of 0.001 (expressed also as 10^{-3}) is represented by a pH of 3; a concentration of 0.0001 (10^{-4}), with 1/10th the concentration of the former, is represented by a pH of 4.

Acidity increases going down the pH scale and its converse, alkalinity, increases moving up the scale. A pH of 7 is neutral; above

7, conditions are alkaline, and below 7, conditions are acidic. For reference, the pH of lemon juice is about 2 and that of household ammonia about 11. Soils generally range from pH 3 to 10, and most cultivated plants enjoy slightly acidic conditions with a pH around 6.5. Pin oak is among those plants, along with gardenia, blueberry, azalea, and rhododendron, that languish if soil acidity becomes less acidic than the very acidic (for soils, that is) pH of 4.0 to 5.5.

The effect of soil pH on plants is indirect, but far-reaching. Those pin oak leaves, for example: the yellowing indicates iron deficiency, a condition arising not usually from lack of iron in the soil but from insufficient acidity to put iron into a form that pin oaks can absorb. Plant nutrients vary in how soil pH affects their availability. Pin oak and company aside, the slightly acidic pH optimum for most plants affords them good access to all nutrients.

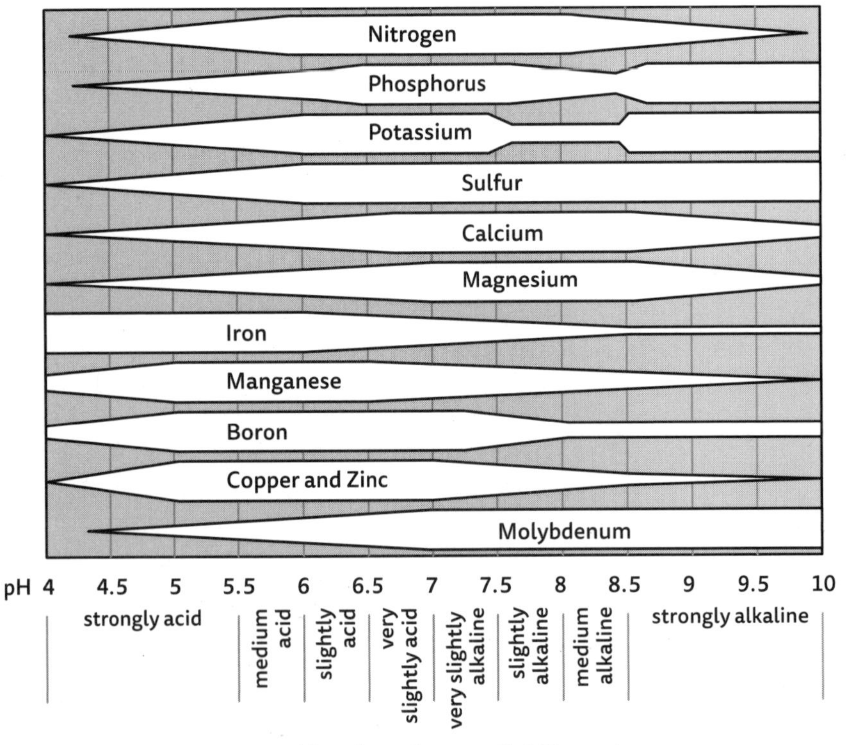

pH and nutrient availability.
Thickness of white area indicates a nutrient's availability.

Plant poisoning is the darker side of soil pH. Low pH can render manganese, a plant nutrient, so available that it becomes toxic; geraniums are particularly sensitive, showing their discomfort with yellowing, dying, and brown-flecked leaves. Low pH also liberates aluminum, which is not a plant nutrient, in amounts that can stunt root growth and interfere with plant uptake of nutrients. At the other end of the pH scale, the plant nutrient molybdenum becomes available in toxic amounts.

Soil pH also affects soil-dwelling organisms, whose well-being, in turn, affects plants. The same slightly acidic conditions enjoyed by most plants are also what earthworms enjoy, as do most beneficial microorganisms. Manipulating the soil pH combats certain diseases, with pH 5.2 or less suppressing common scab of potatoes and, to a lesser extent, a pH of 5.5 or lower suppressing damping-off disease of seedlings. Adjusting the soil pH to values above 7.2, on the other hand, controls clubroot disease of the cabbage family.

When I planted blueberries, I knew my soil needed to be acidified; I also knew that just sprinkling, or wantonly throwing, sulfur or limestone on the ground was not going to do the trick. For starters, I needed to know the initial pH of my soil. A simple soil test for acidity can be done at home or by a soil testing laboratory. Next, I needed to know my soil's texture, that is, whether it was mostly clay, mostly sand, or something in between. The charged surfaces of most clays and other small mineral particles make them more resistant to pH changes than the uncharged surfaces of sand particles, so more material is needed for pH changes of clays than of sands. The final bit of information needed was the pH to which I wanted to transform the soil. Blueberries demand very acidic soils (pH 4.0 to 5.5).

Generally, limestone is used to make soils more alkaline (higher pH) and sulfur is used to make soils more acidic (lower pH). Both these materials are mined from naturally occurring deposits. About five pounds of limestone are needed to increase the pH of a hundred square feet of sandy soil by one unit, while about twelve pounds would be needed if the soil were high in clay. The equivalent amounts of sulfur needed to lower the pH by one unit would be a pound in a sand and three pounds in a clay. My clay loam's initial pH was about 6.5, so I needed 3 pounds of sulfur to lower that pH to 5.5, the upper limit enjoyed by blueberries.

Limestone, as mined, is either relatively pure calcium carbonate or, if dolomitic limestone, a mix of magnesium and calcium carbonate. Dolomitic limestone has the advantages, besides neutralizing acidity and adding calcium to the soil, of also adding magnesium and, pound for pound, neutralizing more acidity than plain limestone. Magnesium levels are naturally low in soils of the eastern and northwestern United States and highest in the upper Midwest and the Southwest.

Pine needles or oak leaves are often touted for their ability to acidify a soil. Not so. When organic materials are added to a soil, there is a short-lived, temporary drop in pH, but eventually the pH creeps back up to neutral. Not that these organic materials, like any other organic materials, aren't beneficial to the soil. Just that they should not be relied upon to change acidity.

Neither limestone nor sulfur is soluble in water, so they need to be mixed thoroughly into the top six inches of soil when a quick change in pH is needed. To maintain pH in the proper range rather than quickly change the pH, these materials can be spread over the ground to work their goodness down into the soil over time. For my blueberries, I spread the sulfur over the ground but also mixed it in with the soil in each planting hole; by the time the blueberry roots spread beyond the planting hole, the sulfur (in oxidized form) there would have seeped down into the ground. Both limestone and sulfur are available either powdered or pelletized, with the latter being easier to spread uniformly and causing less of a health hazard from dust. (Elemental sulfur is also sold as a fungicide, for which use it is finely powdered and much more expensive than pelletized sulfur.)

Soil acidity adjusted to the correct range for whatever plants are being grown does not give license to forget about soil pH. Maintaining the correct soil pH is an ongoing task, especially in the naturally acidic soils of the East and the Pacific Northwest, where rainfall leaches out calcium and other alkaline-forming elements. Acid rain, sulfur from the burning of coal, and plant uptake of alkaline-forming elements takes these soils further on the road to acidity. Naturally alkaline soils keep shifting in that direction because of the rock minerals from which they are formed, so much so, in some cases, that acidifying them is unfeasible. Even some chemical (synthetic) fertilizers can shift soil pH over time, with materials such as ammonium sulfate and ammonium nitrate pushing the pH down and calcium or potassium nitrate pushing values up. Hence the necessity for regular additions of limestone or sulfur.

Over the years, I have, admittedly, paid less attention to the pH of the soil at my blueberries' feet. That's because, for many years now, I've annually blanketed the ground beneath those bushes with a 3 inch mulch of some weed-free organic material such as wood shavings, leaves, or wood chips. The soil there is consistently moist and well-aerated because it's become so rich in organic matter whose buffering capacity also allows for more wiggle room in pH.

As for that sickly pin oak, the problem commonly occurs on trees planted near walls, where the soil has been made too alkaline from mortar spilled and plaster or wallboard buried there during construction. With a young tree, it may still be possible to mix sulfur into the soil to lower the pH. For an established tree, I'd just spread the sulfur over the ground and wait longer for effect.

On my ostensibly occult practice which turns out to be good gardening

I wonder if my neighbors suspect that I am engaging in some sort of occult ritual as I take my biweekly winter rounds through the garden followed by puffs of grey "smoke." Perhaps I'm entreating tiny gnomes living within the soil to keep weeds at bay next season? Or begging garden gremlins to make my soil fertile? No, and again no! I'm merely spreading wood ashes about.

Spreading wood ashes.

I must be careful with my terminology: I am not disposing of wood ashes; I am fertilizing my soil with wood ashes. Wood ash is a rich source of potassium, a nutrient required by plants in amounts second only to nitrogen. Potassium helps build strong stems and helps plants resist disease. Potassium also regulates the opening and closing of the tiny pores (stomates) in leaves, through which gases pass for photosynthesis.

The close connection between potassium and wood ash is reflected in a traditional source of, and the root of the word, potassium—"potash." Potassium compounds once were extracted by mixing wood ashes with water in iron pots. Go to a nursery and ask for a potash fertilizer, and you will be handed a bag of potassium sulfate (sulfate of potash), potassium chloride (muriate of potash), greensand (a mined mineral), or—if the time was a century ago—wood ash. "Potash" technically means potassium oxide, but sometimes also refers to potassium carbonate or hydroxide.

Wood ash contains from one to about ten percent potash. (Since "potash" includes the weight of potassium and oxygen, the raw potassium concentration is always less than the potash concentration; 17 percent less, to be precise.) Ashes from hardwood trees are at the top of this range and softwood ashes are at the low end. I always store my ashes in a metal can under cover of my garage/barn, because rain will leach much of the nutriment from ashes.

I am not overly precise in my spreading of these ashes. Because I burn 2 to 3 cords of wood each winter, I have plenty of wood ash, so I disperse it over my whole property, aiming to spread no more than two pounds of ashes over every hundred square feet.

Wood ash is good stuff, but is not to be overused. Plants need adequate, not excessive, potassium. Too much potassium upsets the balance of other nutrients in a plant. Wood ashes also make the soil more alkaline, so I never spread them beneath my blueberry or rhododendron bushes, nor should wood ashes be used on azaleas, pin oaks, mountain laurels, or other plants that enjoy very acidic soils. Wood ash can be substituted for limestone but the pound for pound equivalence varies depending on the kinds of wood that created that ash. Its effect on soil acidity is generally one-fifth to one-half that of limestone.

I do not dump—whoops, spread—all my wood ashes on the garden and lawn. Some I save for use in potting soils, mixed in at the rate of a half-cup of ashes per gallon of mix (except in potting mixes destined for acid-loving plants, of course).

And some I sift, then save, to use in the garden during the summer for pest control. Plants sprinkled with dry ashes become unpalatable to rabbits, bean beetles, and onion and cabbage maggots. A thick line of dry ashes on the soil becomes a barrier to slugs, until washed by rain. Conversely, rain washes into the soil the alkalinity from wood ashes spread over the ground, killing cutworms. Cucumber beetles are said to be repelled (or killed?) by a spray made from a handful each of wood ash and hydrated lime mixed into two gallons of water, then sprayed on the leaves. Squash bugs are allegedly repelled by a sprinkling of wood ashes into which has been mixed turpentine, at the rate of one tablespoon turpentine per gallon of wood ash. And a paste of wood ashes and water on the trunk of a peach tree is supposed to keep borers at bay.

Squash bugs have never been a serious problem in my garden and my cabbages rarely get maggots, so I can't personally vouch for all these pest-deterring and killing properties of wood ashes. But wood ashes are nontoxic and the above suggestions come from reputable sources, so the treatments are worth a try when needed. The main drawback to using wood ashes for pest control is their need for reapplication following rain. Also, the ashes should not be used on seedlings, or they will be "burned."

Perhaps there is some sort of gnome or gremlin in these ashes that imparts to them such a myriad of uses?

How I manage to tame nitrogen's comings and goings for my plants

Nitrogen is at the same time the most needed, the most abundant, and the most fleeting of plant nutrients. And as much as nitrogen is needed by plants, there is no good soil test for it.

Nitrogen is a key component of proteins and, most visibly, of plant chlorophyll. Chlorophyll, coupled with sunlight, fuels plant growth so one symptom of a plant that is hungry for nitrogen is slow growth. (Watch out though: diseases, cold weather, and hunger for other nutrients can also be responsible for slow growth.) Chlorophyll is what makes leaves look green so another symptom of nitrogen deficiency is pale green or yellow leaves. Nitrogen can be shuffled about within a plant, so it is the oldest leaves, which sacrifice their nitrogen to the youngest leaves, that are the first to grow pale.

Nitrogen can be overdone. Plants gorging on this nutrient are overly succulent and barely able to hold themselves up. Overfeeding of nitrogen can also lead to pest problems, most notably, aphids.

Air is the ultimate source of nitrogen but that eighty pounds of nitrogen hovering over each hundred square feet of my garden does nothing to assuage my plants' hunger for the stuff. Atmospheric nitrogen is a gas with two nitrogen atoms bound together into a single molecule; plants, however, mostly take up nitrogen only when dissolved in water as nitrate (nitrogen combined with oxygen) or ammonium (nitrogen combined with hydrogen). Lightning and rain together can convert and bring down a bit of that atmospheric nitrogen into something that plants can use, but only about two-hundredths of a pound per hundred square feet.

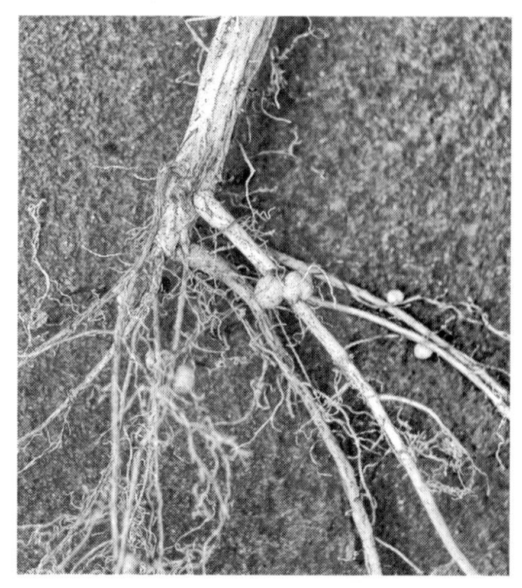

Soybean roots with nitrogen-fixing nodules housing bacteria.

Far more atmospheric nitrogen is made available to plants by what is called "fixation." Biological fixation, by bacteria and other microorganisms, can put enough nitrogen into the soil to feed plants all they need, and played a significant role in doing so, before chemical ("synthetic") nitrogen fertilizers came into widespread usage. I like to see biological fixation for myself as I'm pulling up old bean plants. The pea-size nodules I see there are home to *Rhizobium* bacteria which take nitrogen from the air and feed it to the plants. *Rhizobium* species for specific kinds of peas or beans are available commercially. I dusted the appropriate inoculant for the beans and peas when I first planted them in the garden's beginnings. These friendly bacteria persist in the soil, obviating the need to inoculate for subsequent plantings.

A healthy population of *Rhizobium* infecting my bean roots is also a sign that I'm not over-fertilizing with nitrogen, which suppresses the infection.

In traditional and organic agriculture, legumes (which include peas and beans) are periodically grown and then worked into the soil to release their fixed nitrogen to subsequent crops, or else their nitrogen-rich seeds are ground up and spread as fertilizer. Biological nitrogen fixation also occurs in rice paddies and in association with the roots of such shrubs as alder and buckthorn. As animals feed on plants, legume or otherwise, some of the fodders' nitrogen is concentrated and then excreted in their manure, another organic source of nitrogen.

The nitrogen in organic materials is bound up and unavailable to plants. But as microorganisms get to work on organic materials, breaking them down into simpler and simpler components, nitrogen is released in a form plants can "eat." The stepwise reaction goes from proteins to amino acids, and then to ammonium, which plants can use, and then to nitrate, which plants also can use.

Proteins ⇒ Amino acids, Amides, and Amines ⇒ Ammonium ⇒ Nitrate

Chemical nitrogen fixation takes place in factories via an energy intensive process of combining air with natural gas to produce chemical fertilizers. Worldwide, chemically fixed nitrogen amounts to somewhat less than a third of that fixed biologically. One other, smaller source of chemically fixed nitrogen is the exhaust gases spewing out our tailpipes and smokestacks and eventually brought down to the soil by rain.

Although roots finally imbibe most nitrogen as nitrate and/or ammonium, all nitrogen sources are not equal to plants. If I were to go to my local garden center to purchase a chemical fertilizer, the nitrogen in that bag would be already in one or both of the two available forms, immediately ready for plants to imbibe. Used to excess, chemical fertilizers, which are salts in the general sense, suck water from roots in the same way that salty potato chips dry out our lips. Hence, the recommendation to apply these fertilizers repeatedly through the season rather than in a single, full dose.

How soon the nitrogen in organic sources of nitrogen becomes available to plants depends on how much organic matter is present and how fast it decomposes, which, in turn, depends on how tightly the nitrogen is initially bound up, and how much warmth and moisture spur microbial activity. Increasing warmth and moisture also spur plant growth. Voila! The organic nitrogen is made available to plants in sync with their needs. Some advance planning is needed when using organic nitrogen fertilizers so that their nitrogen is ready when plants need it.

Elegant Plan for Acidic Soils

Plants that thrive in very acidic soils, such as rhododendrons and azaleas, prefer their nitrogen in the form of ammonium rather than as nitrate. Local garden centers stock fertilizers tailored to the needs of these plants, fertilizers such as Miracid and Holly-tone whose source of nitrogen is ammonium.

Most organic fertilizers naturally fill the bill—for acid-loving as well as for any other plant. The last step in the conversion of complex nitrogen compounds in organic materials to simple, plant-ready compounds is ammonium to nitrate. This is the responsibility of nitrifying bacteria—except they are absent in very acidic soils. So the nitrogen conversion stops at the second to last step, at ammonium, which is just the kind of nitrogen that acid-loving plants like to eat. That same organic fertilizer could also be used in a more neutral soil, in which the nitrogen reaction proceeds on to nitrate.

(Both ammonium and nitrate are fine nitrogen foods for most plants. Ammonium takes less energy to take in and assimilate but, except for acid-loving plants, can be toxic in excess. Nitrate is more easily stored in plants.)

Nitrogen can be lost from a soil in about as many ways as it can be added. Nitrate nitrogen is very water soluble so readily leaches down beyond the reach of plant roots (potentially polluting groundwater) following heavy rain or irrigation. Hence, the recommendation to apply chemical nitrogen fertilizer in spring and summer, when roots are ready to use it. Ammonium nitrogen is less subject to leaching because its positive charge binds to soil particles. Organic nitrogen fertilizers are even less likely to leach because they're initially insoluble and even when warm, moist weather finally makes them soluble, they transmute first into that less likely to leach ammonium form.

Nitrogen in various forms also can waft out of the soil as a gas. This happens under waterlogged conditions, which anyway are hostile to most plants. Or, if a nitrogen-rich manure or ammonium fertilizer is left on the surface of the ground, especially ground to which limestone has been recently applied, the smell of ammonia will be in the air, a sign that nitrogen is wafting away.

All nitrogen's comings and goings are what make testing for it at a given point in time of dubious value. The amount of organic matter in the soil, which can be accurately measured by a soil test, does give some indication of a soil's potential to supply nitrogen to plants. Each percentage of organic matter in a hundred square feet of soil contains about two pounds of nitrogen, one to four percent of which is released as nitrate or ammonium every year. So if a soil contains three percent organic matter and two percent of the nitrogen will be mineralized each year, a rough estimate would be for a natural release of 0.12 pound of nitrogen from that organic matter each year ($2 \times 3 \times .02$). This is slightly more than half the two-tenths of a pound per hundred square feet generally needed by vegetable plants.

If the soil was higher in organic matter and/or warmer and moister conditions increased the rate of decomposition, that organic matter could completely fulfill plants' nitrogen needs. I'm proud to say that the level of organic matter in my soil is 15%, which computes, roughly, to a release of ($2 \times 15 \times .02$) or 0.6 pounds of nitrogen per hundred square feet per year—more than enough for my intensively planted vegetables. (I have wondered whether it might be too much although it doesn't seem so, at least judging from the good nodulation on the roots of my peas and beans. Also, that figure of two pounds of nitrogen released per year per percent of nitrogen might apply to tilled soils; my soils are never tilled, so nitrogen release in my soil may be less—but, judging from plant performance, totally adequate.)

Soil test recommendations for nitrogen for a vegetable garden hedge by assuming plants need to be fed that whole two-tenths of a pound per hundred square feet of nitrogen each year. That's actual nitrogen, expressed as elemental nitrogen, N (which doesn't exist as such in nature); how much fertilizer to apply depends on how much nitrogen a fertilizer contains. That would be two pounds of a fertilizer that is ten percent nitrogen, four pounds of a fertilizer that is five percent nitrogen, etc.

I tailor my nitrogen feeding to the plants I am growing. Garden vegetables are nitrogen hungry plants. I used to apply plenty of compost (one form of organic matter) along with a once a year sprinkling of soybean meal, which is about 7% organic nitrogen. My computations showed that a one-inch depth of compost could supply all the needed nitrogen for a year, so that's all the beds now get each year. In my flower bed of coneflowers, liatris, and other plants that prefer a relatively spare diet, I do nothing more than mulch annually with leaves or wood chips, either of which decomposes over time to release some nitrogen into the soil. Most important is keeping a watchful eye on plants to best assess their needs and then tailor feeding as appropriate.

Even without squealing like hungry pigs, my plants can tell me if they're hungry, and for what

Like you and me, plants need a variety of nutrients to stay healthy. Feeding plants only nitrogen won't do. Fifteen mineral elements are recognized as being essential to plants (13 or 14 by some counts), in addition to carbon, hydrogen, and oxygen, which come from CO_2 and water, and four—silicon, sodium, cobalt, and selenium—that are deemed beneficial but not essential. "Essential," here, means the element must be needed for a plant to complete its life cycle, it must be directly involved in metabolism, another element cannot substitute for it, and it's needed for a wide range of plants. Those essential mineral elements are nitrogen, phosphorus, potassium, calcium, magnesium, sulfur, boron, chlorine, iron, manganese, zinc, copper, molybdenum, and nickel.

Three numbers emblazoned on bags of fertilizer indicate the contained percentages of nitrogen, phosphorus, and potassium, the "primary elements" needed in greatest amounts by plants.

(Actually, those three numbers do not indicate exact percentages. Rather than being expressed in terms of the simple elements, the amounts of phosphorous and, as mentioned previously, potassium, are usually expressed in the traditional manner, as percentages of the oxides "phosphoric acid" and "potash." Fertilizer manufacturers are reluctant to break tradition and list actual percentages because then a fertilizer such as 1-2-1 would become 1-.9-.8, leaving people to believe—erroneously—that they were getting less for their money.)

Any plant being fed only 1-2-1, 10-10-10, or any other fertilizer containing only nitrogen, phosphorus, and potassium would, obviously become malnourished. The "secondary elements," calcium, magnesium, and sulfur, often hitchhike along as carriers in a fertilizer mix or are added when liming a soil to make it less acidic. The remaining elements, the "micronutrients," are no less important than the aforementioned macronutrients, but are needed by plants in only minuscule amounts. Micronutrients originate mostly in soil minerals so are most likely to be lacking when plants are grown without real soil. These conditions might exist when plants are grown hydroponically or in soilless potting mixes (most store-bought potting mixes are soilless). Certain regions are predisposed to certain micronutrient deficiencies, such as boron in sandy soils in regions that experience moderate or heavier rainfall.

Micronutrients are fickle, and just because they are present in a soil doesn't mean that plants are able to use them. Soil drainage, acidity, and organic matter are some factors influencing micronutrient availability, with good drainage, slight acidity, and ample organic matter generally keeping plants well-nourished in these nutrients. Manganese, for instance, becomes toxic to sensitive plants, such as tomatoes, in waterlogged soils. Steam-sterilization of potting soils to eliminate soil-borne diseases can also bring on manganese toxicity. The micronutrients chlorine and boron often accumulate naturally to toxic levels in arid regions.

Excess or deficiency of one nutrient can also cause micronutrient problems. Overfertilization with phosphorus fertilizer, for instance, leads to zinc or iron deficiency, and overfertilization with zinc leads to copper deficiency.

Iron deficiency of blueberry leaf.

Hungry plants don't squeal like hungry pigs, but can show when they are very hungry, with poor or distorted growth, and with leaf discolorations characteristic for whatever nutrient is lacking. Of course, insects, diseases, environmental conditions, and imbalances of major nutrients could also be the culprit.

Laboratory testing is the surest way to assess the micronutrient status of a soil. That laboratory could be at home, with a home testing kit, or a state or private laboratory. The test itself is done on either the soil or the plants themselves. An important part of soil testing is getting a representative sample. Equally important, for a soil or plant test, is avoiding contamination of the sample; use clean stainless steel, glass, or plastic utensils and containers.

The usual way to alleviate deficiencies of specific micronutrients, or a spectrum of them, is with soluble, inorganic salts ("salts" in the general sense, not necessarily sodium chloride), with synthetic chelates, or with frits. The salts or the chelates can be added to the soil or, for quick, emergency response, sprayed directly on leaves. One problem with soluble salts is that they are quickly immobilized in the soil. Chelates avoid this problem by enclosing the micronutrient element in a chemical ring structure that keeps it soluble and available to plants. Frits are micronutrient salts that have been fused with silicates to make a glass-like material; they are insoluble so must be mixed into soils, and perform best in acidic soils. Salts, chelates, or frits must be applied with care because of the narrow window between deficiency and toxicity with some micronutrients.

I take the low-tech route to avoiding micronutrient deficiencies in the first place. Keeping my soil well supplied with plenty of organic materials—especially compost—derived from a variety of sources keeps my plants well nourished with micronutrients. The wood chips, leaves, hay, and compost that I am always adding to my soil become part of the soil's organic matter pool, which has natural chelates that, just like the synthetic chelates, help keep micronutrients available for plant use. Like organic fertilizers, these organic materials also themselves contain a spectrum of plant nutrients, including micronutrients. In addition, my garden plants get micronutrients from peels of Florida oranges, pits of California avocados, peelings of Canadian carrots, and other kitchen scraps which, along with manure, hay, and other materials, make their way into my compost piles and thence to my garden.

Blossom end rot of tomato, due to calcium deficiency of fruits.

To round out the spectrum of nutrients available to my plants, I occasionally sprinkle powdered kelp on the compost pile or the ground, although its value is dubious, considering the wide spectrum of feedstuffs added to my compost. Still, I figure we all originally came from the sea so the sea might have all the nutrients we need. Not very scientific, but quickly applied and cheap insurance.

Nutrient Deficiency Symptoms

Nitrogen: The oldest leaves are affected first, and they turn pale green and then yellow. Yellowing usually begins first at the tips of the leaves, but in corn the midrib of the leaf is the first part to turn yellow.

Phosphorus: Here again, the oldest leaves are the first to be affected. The leaves take on a reddish or purplish cast.

Potassium: The oldest leaves die beginning at their tips and proceeding, in contrast to nitrogen deficiency, along the margins of the leaves.

Calcium: Calcium becomes fixed within plants, so when plants become deficient, it stays put and the tips of the plant can no longer grow. Calcium deficiency causes the bottoms of tomato fruits to turn black, a disorder called "blossom end rot."

Magnesium: Magnesium is mobile in plants, so older leaves are affected first. These leaves turn yellow in between their veins.

Zinc: A deficiency causes shortening of the distance from one leaf to the next, so rosetting occurs; another symptom is yellowing between the veins of young leaves.

Sulfur: Deficiency is rare, thanks to "acid rain."

Iron: Insufficient iron results in yellowing between the veins of young leaves.

Manganese: A deficiency results in yellowing between the veins of young leaves, but with a gradation of darker and lighter color rather than the sharp demarcation seen with iron deficiency.

Copper: A deficiency results in stunted growth and dark, green leaves.

Boron: A deficiency causes death of terminal growth, which causes lateral buds to develop, so a "witches' broom" develops; leaves might also become thickened, curled, and chlorotic.

Molybdenum: In legumes, stunting and lack of vigor result with leaves turning pale, as with nitrogen deficiency.

Chlorine: Deficiency results in wilting followed by chlorosis; underground, there is excessive branching of lateral roots.

FLOWERING AND FRUITING

Sex is introduced and its sometime importance is emphasized.

Come winter holiday season, I might deck my halls with boughs of homegrown holly, but unless I planned ahead, those boughs could lack red berries. And that leads to some frank talk about sex.

A holly berry, like any other fruit, is a mature ovary, which is a home for a seed or seeds. Seeds are what stimulate development of a fruit, but seeds themselves usually can't get started without sex.

Sex happens in plants when pollen from the male part of a flower lands on the female part of a flower, called the stigma, and then grows a pollen tube down the style, which is attached to the stigma, to reach and fertilize an egg. Effective fertilization only occurs between flowers of the same species, sometimes between species, and, rarely, between genera. So crabapple flowers can fertilize crabapple flowers, but can't fertilize pear flowers. Depending on the plant, pollen is transferred by wind (known, euphoniously, as anemophily), insects (especially bees), or when the mere opening of a flower causes stamens to rub against pistils. The product of successful pollination and fertilization is a seed, the development of which stimulates the surrounding floral part to swell to become a fruit.

Why am I so concerned with holly's sex life? After all, I don't give sex a second thought when growing tomatoes. I plant whatever varieties I want and then reap plenty of swollen ovaries...er, fruits...as well as, incidentally, seeds.

Holly is special because its pollen is borne on flowers that are strictly male and its eggs are contained within flowers that are strictly female. Each tomato flower, in contrast, has both male and female parts, so can take care of itself. Similarly self-sufficient are rose flowers, peach flowers, sunflowers, and the flowers of many other plants. Plants whose flowers house both male and female parts are called, botanically, "perfect" flowers.

Holly is not alone in having single sex flowers, known botanically as "imperfect" flowers. Many nut trees, for example, share this trait.

Holly goes one step further, sexually, with whole plants being either male or female, a trait shared by ash and persimmon trees, among others. Such a situation encourages species diversity by mandating cross-pollination among different plants. Nut trees, whose male and female flowers are borne on the same plant, achieve the same infidelity with biochemical or physical barriers, or different bloom times that prevent male flowers from

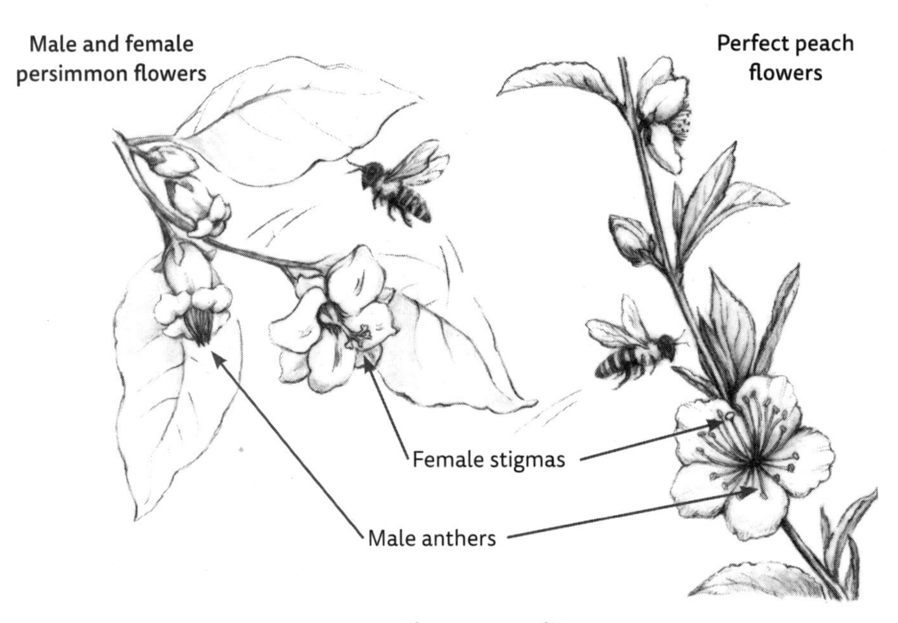

Flower sexuality.

pollinating female flowers on the same plant. Even some perfect flowers, such as apple blossoms, have biochemical or physical barriers preventing self-pollination, so two different varieties of apple are needed to get fruit; both trees bear fruit. And some perfect flowers, such as blueberry, can pollinate themselves, but more and bigger berries result when pollen is exchanged between different varieties.

The long and the short of all this with holly is that an all-male holly tree or bush is needed if I'm going to deck my halls during winter with (berried) boughs of holly from my all-female holly tree or bush. A male plant, all leaves and no berries, isn't as showy as a female, so it's fortunate that a single male can sire a half-dozen or so females. The males, like the females, do bear flowers, but neither male nor female holly flowers are worth a second look unless you want to peer at them closely to determine the sex of a plant.

No need, perhaps, to deliberately plant a male holly in order to get a female to yield berries. Suitable pollen conceivably could come, instead, from wild or neighbors' trees (perhaps a male kindly offered to plant in your neighbor's yard, heh, heh), or from some flowering stems clipped from a male holly, then set next to the females and kept fresh with their bases plunked into a bucket of water, or a single male stem could be grafted onto a female plant. The same could be done for an apple tree or any other tree needing a pollinator when there's none nearby and only one tree is wanted.

I chose to plant one male to sire my five female hollies. Except that the nursery that shipped me the plants mislabeled the male. So, after about ten years of waiting for berries, I purchased a new male—from a different nursery.

Adding to their sex problems, or rather, our problems with their sex life, hollies are not all that promiscuous. A few different species supply red-berried boughs—notably American holly, English holly, and Meserve holly—but, generally, each keeps fidelity to its own species. (An exception is that English holly can pollinate Meserve holly, which is a hybrid offspring of the English species.) Further compounding hollies' sex problems, some males within a species fail to adequately pollinate some females within the same species because their bloom times do not overlap.

If a male holly is needed, breeders have come up with a number of superior male varieties whose genders are obvious from their names: 'Blue Prince' and 'Blue Boy' Meserve hollies, and 'Jersey Knight' American holly are examples. These males, as you might guess, are particularly good mates for the varieties named, respectively, 'Blue Princess', 'Blue Girl', and 'Jersey Princess'.

Some plants don't bother at all with this sexual intrigue. Many varieties of persimmon and fig, for example, are "parthenocarpic" (meaning, literally, "virgin fruit"), that is, their fruits develop without the need for sex. In such cases, not only is a male plant or branch unnecessary, but the parthenocarpic fruits also are seedless. And many plants—from maples to marigolds—don't interest us for their fruit, in which cases, we can forget about sex. Their limbs, their leaves, or their flowers are sufficient to satisfy us.

In which I make right the products of plants' sexual excesses

Snowballs of bloom followed by miniature fruits tell of branches to be bowed by the weight of large and luscious peaches, apples, pears, and plums. Large and luscious, that is, if I pluck some of the fruits off while they are still small.

Removing a portion of the fruits (fruit thinning, as this operation is called) can direct more of a tree's attention to those fruits that remain. The goal of the tree, after all, is to reproduce and spread its progeny by putting its energy into maturing as many seeds as possible. This means ripening lots of fruit, even if they do end up small and not as tasty as they could be. What we humans want, though, are large and luscious fruits, the result of coaxing the tree to put more energy—which means, mostly, sugars produced by photosynthesis—into fewer of them. For perfect ripening, for example, a single apple needs the resources of 20 to 40 healthy leaves, or a dozen in the case of more energy-efficient dwarf trees. All fruits suffer in quality if the leaf to fruit ratio drops too low.

Another reason you might see me out in the garden in spring pinching some fruitlets off my fruit trees is to make them bear more consistently. My 'Macoun' apple knows no moderation: it wants to ply me with oodles of fruit one year, then starves me the next, feast alternating with famine year after year. But I want to bite into these delectable 'Macoun' apples every year.

'Macoun' and some other fruits get into this feast and famine habit because seeds within the fruits produce a hormone or hormones that suppress flower bud initiation. Most fruit trees develop flower buds the year before the buds actually open, so abundant fruit one year means many seeds, less flower bud formation, and, hence, less fruit the following year. Thinning out fruits reins in the tendency towards biennial bearing.

A straightforward benefit of fruit thinning is, of course, nothing more than physics: less weight on a branch. A too-heavy fruit load often breaks branches.

Putting space between fruits also has some effect on pests. Codling moth caterpillars—the half "worm" (caterpillar, actually) left when you take a big bite into an infested apple—prefer to tunnel into fruits that are touching each other. Putting space between fruits also lets each one better bask in air and sunlight, both of which hasten drying and so make it harder for diseases to gain foothold.

Most fruit trees require annual pruning and as I prune to open a tree up to light and air, and to control its size, I am indeed removing potential fruits and seeds because I'm reducing the number of flowers before they even open. In addition, shortening branches puts any fruits that do set closer to the trunk where, with less leverage, limb breakage is less likely. Pruning alone, though, is generally not enough to get fruit trees that bear large fruits to ripen their fruits to perfection or to get them out of a bad habit of biennial bearing—which is why I am also reaching in among my fruit trees in May and June, pinching off excess fruitlets.

So much for theory; the hard part is actually bringing oneself to remove promising little fruitlets. I grit my teeth and snap them off with my fingers or use a small clippers, taking care not to damage the knobby little stem to which an apple, pear, or plum stalk is attached. That swollen stem is the origin of flowers and fruits in years to come. Gardeners with larger trees or less patience might resort to thinning fruits by blasting branches with a stream of water, batting branches with a piece of hose slipped over the end of a broom handle, or brushing branches with a stiff brush. Commercial fruit farmers have chemical sprays—some contain synthetic hormones—that accomplish this same task on a larger scale and with more physical delicacy.

No matter how thinning is done, the fruits to leave are those that are largest and healthiest. Many of the small or damaged fruits will eventually fall off anyway. I leave a space between fruitlets of two to three times the diameter of the mature fruit. The earlier fruit thinning is done, the better, especially with apples and less so with peaches, then pears. Fortunately, thinning is

Hand thinning of apple fruitlets.

unnecessary with smaller fruits, such as cherries and European plums. How tedious that would be!

In addition to, or in the absence of, human intervention, fruit thinning also occurs naturally. Fruit trees are determined to set as many fruits as possible right after blossoming to increase the likelihood of more making it through any post-bloom frosts, but these plants do have some sense. A few weeks after bloom, once the weather has settled, the trees realize how taxing it would be to mature all those fruits, so they shed some. Not enough of them, in our opinion.

Describing the importance of night for coaxing blossoms, and a gardener's trickery

As far as I am concerned, the first hint of spring is in the air in early January. Yes, really. No matter what the temperature is outside through January and on into the equally cold month of February, the days are then growing longer—more spring-like. And along with lengthening days, the sun rises higher and higher in the sky, increasing in intensity.

The effects of increasing illumination are first noticeable on houseplants. What a change from autumn, when growth is sluggish or grinds to a halt. With so little light, even the warm temperatures of an Indian summer can't spur growth from trees and shrubs outdoors.

It's not only the intensity of light that affects plants, but also its daily duration. Back in the 1920s, two scientists, W. W. Garner and H. A. Allard, found that the only way they could get their Mammoth Maryland variety of tobacco plants to flower was by exposing them to short days. They could either grow the plants in pots in a greenhouse and wait until December, or they could make the plants flower in spring or summer by covering them for part of each day to exclude light.

The response of plants to day length is called "photoperiodism." Besides Garner and Allard's tobacco plant, a number of other plants—including corn (some varieties), primrose, freesia, 'mums, and aster—flower and fruit only if they are first exposed to a period of short days. In other experiments, Garner and Allard found that soybeans would flower at the same time in late summer irrespective of when they were planted in spring. Beets, radishes, spinach, gladiolus, and lettuce are among those plants that only flower after a period of long days. A third group of plants, including dandelion, buckwheat, sunflower, beans, and many tropicals, bloom irrespective of day length.

Subsequent research revealed that photoperiodic plants are not really responding to the length of the days but, rather, to the length of the nights. Phytochrome, a blue-green pigment found in plants, is sensitive to infra-red radiation, which is present in light and converts phytochrome to its active form. In dark, or with exposure to far-red light (the part of the electromagnetic spectrum just beyond the red that we can see but shorter in wavelength than infra-red), the active form converts to the native form. The shorter the night, the less conversion takes place. Despite these later findings that night length is really what makes plants respond, the original terms "short day" and "long day" have stuck.

"Long day" onions bulbing up with summer's long days.

Photoperiodism is of practical use to us gardeners, and the difference between saying "short day" and "long night" is more than semantic. For instance, a poinsettia will bloom again each winter holiday season only if it gets two months of daily exposure to 14 or more hours of darkness, beginning in September. Total darkness. Even dim light (as little as 2 foot-candles) or a sequence of bright flashes during the dark period makes any short-day plant feel like it has had two short nights rather than one long night.

I pay attention to photoperiod when buying onion seeds or plants for my garden. Onions are photoperiodic with respect to bulbing. In the North, long-day varieties are the ones to grow; they would not bulb up in the South because days there never get long enough, even in summer. Conversely, if I planted a short-day onion here in New York, bulbing would begin before plants would have grown enough leaves to fuel full-size bulbs.

That's also why, when I grow lettuce, spinach, and radish seeds in spring, I make sure to get them in the ground early. Photoperiodism at work here, again. These short-day plants turn bitter or pithy when they send up flower stalks, so I want them harvested before days get too long.

I pay attention, also, to seed depth. Photoperiod-sensitive seeds need light to germinate. Typically, these are small seeds,

Lettuces going to seed with summer's long days.

which makes sense because planted too deep, or buried at all in some cases, a small seed would exhaust its energy reserves before its first shoots broke through the ground to unfurl leaves and start gathering sunlight.

Photoperiod doesn't work alone in prodding plants to grow, flower, or bulb. Temperature also figures in. Spinach, for example, usually flowers when days are 14 or more hours long while the plants are growing, but will also do so following eight hour days if the seeds are first chilled. And plants need to be a certain age before they can flower, which is why I make repeated sowings of lettuce—every two to three weeks—to always have younger plants as temperatures warm into summer. I'll also keep lettuce or spinach cooler by planting them in the temporary shade of trellised cucumbers. Water might also figure into photoperiodic response, which makes a good case for keeping the soil for lettuce and spinach moist by enriching it with organic matter and through timely watering.

Even poinsettia, a tropical plant, flowers in response to temperature as well as photoperiod.

Photoperiod is definitely an important environmental cue—and not only to plants. Photoperiodic effects also initiate mating and other activity in animals as diverse as aphids, codling moths, and many species of birds, fish, and mammals.

In which a small gas molecule has a big effect on flavor

Ethylene is so simple. It's a gas made up of just two atoms of carbon and four atoms of hydrogen. Simple gases are generally not the kinds of molecules that make plant hormones. Plant hormones, like human hormones, are substances having dramatic effects at extremely low concentrations and generally they are complex molecules, and not gases. Nonetheless, ethylene is a plant hormone.

I think of ethylene as I sink my teeth into fresh garden tomatoes in late autumn. Note that I wrote "fresh," not "fresh-picked."

Those tomatoes are picked weeks prior to eating, before the first frost. I pick the pale-green, pink, or red tomatoes that are clinging to the vines, then gently lay them out in single layers on trays in my basement. Every day after picking, the pale fruits deepen in color as they ripen.

It's ethylene that is responsible for this transformation from pale and insipid to red and flavorful. Ethylene is produced naturally in ripening fruits, and its very presence—even at concentrations as low as 0.001 percent—stimulates further ripening. The ethylene given off by ripe apples can be used to hurry along ripening of tomatoes, by placing an apple in a closed bag with the tomatoes. If the fruits are left too long in that bag, ethylene will stimulate ripening which will stimulate more ethylene which will stimulate more ripening which will stimulate more ethylene which will stimulate still more ripening, *ad infinitum*, until what is left is a bag of mushy, rotten fruit. Hence, one rotten apple really can spoil the whole barrel.

Banana, apple, tomato, and avocado are among so-called climacteric fruits, which undergo a burst of respiration and ethylene production as ripening begins. These fruits, if picked sufficiently mature, can, like my tomatoes, ripen following harvest. Soon after their climacteric, these fruits start their decline. (I question whether the color and flavor changes associated with ripening of climacteric fruits after harvest match flavor development of these fruits when allowed to fully ripen on the plants. Ripening on the plant, of course, introduces other variables, such as sunlight, which also influence flavor.)

Citrus, fig, strawberry, and raspberry are examples of nonclimacteric fruits, whose ripening proceeds more calmly. Nonclimacteric fruits will not ripen after they've been harvested. They might soften and sweeten as complex carbohydrates break down into simple sugars, but such changes might be more indicative of incipient rot rather than ripening or flavor enhancement.

Over the centuries, without knowing it, ethylene sometimes has been applied to growing plants or stored fruits. Fruit shippers realized decades ago that oranges could not share a ship's hold

Oiling a fig fruit to speed ripening.

with ripe bananas, or the latter would be overripe at the banana boats' port of arrival. The Chinese used burning incense to hasten fruit ripening (burning gives off a small amount of ethylene). Fig-oiling—putting a drop of oil (often olive oil) on the eye of a nearly ripe fig while it's still on the plant—has been practiced since ancient times to speed fruit ripening. The released ethylene is what does it.

It wasn't until 1934 that a scientist discovered that ethylene is produced in plants and stimulates ripening. These days, ethylene is applied knowingly with a chemical commonly known as Ethrel® (2-chloroethylphosphonic acid). When plants are sprayed with Ethrel®, it changes to ethylene. In an effort to thwart ethylene production—to put the brakes on fruit ripening, to keep harvested, ripe fruit in good condition longer, or to delay fruit drop from a plant—chemical inhibitors, such as 1-Methylcyclopropene (1-MCP) and aminooxyacetic acid (AOA) have also been developed.

Ethylene has effects other than hastening fruit ripening. It also can slow rampant shoot growth—sometimes, in so doing, diverting the plant's energy into making flowers and fruits. Puerto Rican pineapple growers used to build bonfires near their fields to get the plants to fruit. On a windowsill, a potted pineapple plant can be persuaded to flower by enclosing the whole plant in a bag with an apple for a while. Commercial apple trees which are growing only wood sometimes are sprayed early in the season with Ethrel® to get them to settle down and start fruiting.

Fruit ripening and ethylene itself are not the only stimuli to ethylene production in plants. If a leaf is damaged, or even gently rubbed, the cells start emanating ethylene. The same is true if a branch is bent or whipped in the wind. There may be something to the folklore of beating an apple tree to induce it to fruit.

Just as ethylene hastens ripening, then senescence, of fruits, it also hastens senescence of leaves and flowers. A leaf might yellow, then drop, partly in response to ethylene production following insect or disease damage. Ethylene's capacity to hasten leaf drop is used in the nursery industry, when deciduous trees are sprayed with Ethrel® to hasten dormancy and, hence, the time when they can be dug. Cut flowers fade depending on how much ethylene they are producing, both from aging and in response to being cut. Adding a floral preservative, such as silver thiosulfate, to the water in a vase works by putting the brakes on ethylene production.

The tomato fruits in my basement in autumn usually ripen at a satisfactory rate (perhaps hastened by a bit of ethylene from the oil furnace), which is about as fast as they are eaten. When I pick them, I am especially careful to pick only undamaged fruit. Damaged portions of a fruit would produce more ethylene, then ripen and rot too quickly. Some gardeners harvest whole plants before frost, then hang them upside down from rafters. This may be convenient storage but should not affect ripening. Other gardeners wrap each fruit in newspaper, which accelerates ripening by slowing ethylene diffusion from the fruits. I keep my tomatoes unwrapped so I can keep an eye on them, moving the rotten ones to the compost pile and the ripe ones to the kitchen.

Contains a question and an answer: is hybrid always high-bred?

Perusing seed racks and catalogs, you can't help but notice that some seeds are touted as heirlooms and others as hybrids. What are heirlooms, what are hybrids, and are the latter really "high-bred?"

First, let's take a brief return flight to the proverbial birds and bees, in the plant world, with the addition, this time, of the human hand. Suppose that you had two tomato plants, one of which bore fruits that were tasty but sickly red, and the other of which bore fruits that had insipid flavor but were alluringly red. Tomato flowers are perfect and usually self-pollinate naturally, but if you ripped the male parts off a flower on one of the plants, you could dust the other plant's pollen onto that emasculated flower.

The resulting fruit would contain hybrid seeds. With luck, the chromosomes might have segregated and regrouped in such a way that when you sowed those seeds, the resulting plants would bear tomatoes that were both tasty and alluring. A hybrid might also be bred for such qualities as pest resistance, fruit size, ripening season—any and all the things we could like about a plant. Hybrid plants often are more robust than their parents—they have so-called "hybrid vigor."

Seeds won't "come true" (that is, exactly reproduce the parent plant) from hybrid plants. Take the seeds out of a hybrid tomato, such as 'Big Boy', and you will not get 'Big Boy' fruits on those plants next year. So you must purchase seeds of hybrid varieties.

Hybrid seeds yield more uniform plants than do open-pollinated seeds—the name given to seeds from plants that have naturally self- or cross-pollinated for many (plant) generations. The uniformity of hybrid seeds is appealing to the farmer growing ten acres of wholesale cabbage, so that all those cabbages ripen together and with identical heads. As backyard gardeners, though, you and I might not want even six cabbage heads ready to harvest all at once. True, a ripe cabbage head can sit in the ground for a couple of weeks, but not so with other vegetables, such as sweet corn and peas.

The harvest window for sweet corn is only a couple of days, so if you grow hybrid sweet corn, better plan to pick it all at once. Each year, I plant four beds of 'Golden Bantam' sweet corn at about 2 week intervals beginning in mid-May. I likewise plan on harvesting ripe ears from each bed over the course of about 2 weeks, which I can do because 'Golden Bantam', which originated over 100 years ago, is not a hybrid variety.

For the farmer, uniformity does carry with it a risk that the whole field of nearly identical, hybrid plants might also carry some bad traits. Obvious defects would have kept any hybrid from getting beyond the breeder's field, but hidden deficiencies might lurk, waiting for the opportunity to surface. In 1970, thousands and thousands of acres of hybrid corn succumbed to a new strain of southern leaf blight fungus for just this reason. For a gardener, such a loss is disappointing, but not disastrous.

Heirloom seeds, in contrast to hybrid seeds, are from plants whose flowers self-pollinate and have done so for years, making for more or less consistent offspring from year to year. They are called heirlooms because they have naturally self-pollinated and had their seed collected and re-sown, self-pollinated again, and so on for many generations. Plant a seed from one of these plants, and the resulting plant will be pretty much just like its parent, as long as the parent's flowers were not tainted by foreign pollen. The qualities of an open-pollinated variety were not deliberately bred into the plant; the plant was discovered already having these qualities.

An advantage of heirlooms is that you can save the seed for replanting year after year, as long as the flowers self-pollinate. I save seeds of some of each year's best-tasting tomatoes, peppers, and some other vegetables and colorful flowers to plant in the following year's garden. Why? Saving my own seed from year to year saves some money and gives me a bit of independence from seed companies, which, for one reason or another, may discontinue certain varieties. It's also a way to maintain an annual supply of any seeds that seed companies rarely or never offer, such as 'Belgian Giant' tomato, which have been handed down for generations, from parents to children and from neighbor to neighbor.

When it comes to flavor of vegetables and beauty of flowers, hybrid is not always "high-bred." New varieties of sweet peas have beautiful flowers, but they cannot match the intoxicating fragrance of an heirloom variety such as 'Painted Lady', which was introduced nearly two centuries ago. 'Golden Bantam' corn may not be as sweet as newer hybrids, but has much richer, cornier flavor.

That said, one year, on a lark, I planted seeds of the delectable hybrid cherry tomato called 'Sungold'. Surprisingly, the fruits were very close in flavor and appearance to those on neighboring plants raised from commercial 'Sungold' seeds. And thankfully so, because 'Sungold' seed is occasionally in short supply or very expensive from seed companies; if my non-hybrid plants could produce nearly the same fruits year after year, I wouldn't have to even think about buying the seed.

I eventually recognized a halo effect: because I had selected the seeds and I had grown the tomatoes, they seemed to taste better to me. I finally came to my senses: the tomatoes weren't nearly as good. So 'Sungold' returned to my list of seeds for yearly purchase. Except for 'Sungold', though, I rate heirloom tomatoes as being generally tastier than hybrid tomatoes.

When I save seeds from my garden plants, I select them from plants that are healthy. I let fruits or flowers mature, whether they are the dry pods of lettuce or radish plants, the juicy fruits of tomato or cucumber plants, or the dry seed heads of marigolds or zinnias. I can take a few seeds out of tomatoes as I eat them, but a mature cucumber with ripe seeds is unfit to eat. I rinse well, then dry, the seeds from juicy plants. No need to do anything with the dry seeds I pop out of radish pods or rub from the heads of marigolds or daisies, except to pack them away.

What kinds of plants appear the next year will depend on whether the seeds were collected from hybrid plants, and whether the seeds were from flowers that were self-pollinated or cross-pollinated. Cucumbers, for example, have separate male and female flowers (imperfect flowers), so readily cross-pollinate. To propagate heirloom cucumber varieties, cross-pollination must be prevented either by growing them in isolation from other cucumbers or else by bagging and hand-pollinating a few female flowers with male flowers on the same plant. Female flowers are easily distinguished from males even before pollination by their stigmas and, at the base of the petals, swollen ovaries. Without pollination, of course, the whole female flower drops off the plant.

The most predictable outcomes when saving flower and vegetable seeds will be with those taken from non-hybrid plants that have not cross-pollinated or do not do so readily—like heirloom varieties of tomatoes and peppers. Some interesting results can be expected with the others.

Saving Tomato Seeds

If you harvested some particularly good tomatoes this year from store-bought transplants or seeds, you could just plan on buying that same variety to plant again next year. Of course, those transplants or seeds might no longer be available. Saving your own tomato seed gets around this problem, and is easy.

The structure of most tomato flowers causes the flowers to pollinate themselves before they even open, so there's usually no need to worry about seed becoming contaminated by foreign pollen even when different varieties are grown side-by-side. However, flowers of the nearly wild currant tomato do not self-pollinate. If you grow currant tomatoes—and I do recommend them for their delectably sweet, marble-sized fruit—plant them in isolation from other tomato varieties. Do the same if you grow and want to save seed from any of those old-fashioned tomato varieties whose leaves have the same smooth margins as do potato leaves, the so-called potato-leaved types, such as 'Brandywine'. They also easily cross-pollinate.

Saving heirloom tomato seeds.

When it comes time to collect seeds, there is no need to sacrifice the best tomato fruits. What I look for is the plant (if growing more than one of the desired variety) that is healthiest and generally bears fruit typical of that variety. The healthier the plant, the less the possibility of carrying any seed-borne diseases, such as viruses or early blight, on to the next generation.

I choose a fruit, and slice it in half through its "equator." The seed-containing locules are now staring at me, so turning each fruit half upside down over a glass and squeezing it is enough to dribble the seeds out. Then I eat the tomato (this is not a necessary step for seed saving).

To prevent germination while it is in the fruit or as soon as the fruit hits the ground, tomato seeds are surrounded by a gel that contains a germination inhibitor. One way to purge this inhibitor is to let the tomato mush that was squeezed into the glass sit and ferment for 2 or 3 days. That mush soon starts smelling rank, so I prefer another method, which is to add some water to the mush, stir it around, and let it sit for 24 hours. Either way, when the required time is up, I rinse the seeds thoroughly in several changes of water, pouring off waste and nonviable seeds (which float), then strain out the good seeds.

At this point, tomato seeds need to be dried thoroughly and quickly, or else they will sprout. I shake off excess water from the strainer, then dump the seeds onto a plate or a few sheets thickness of newspaper and spread them out. A fan or a sunny, protected spot speeds drying, which should be completed within a couple of days or less.

STEMS AND LEAVES

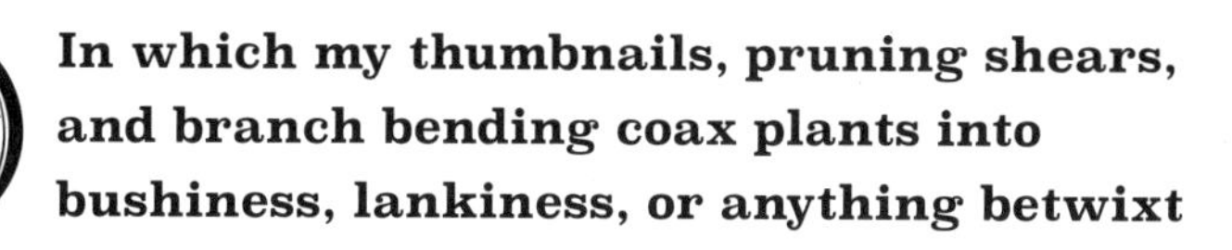

In which my thumbnails, pruning shears, and branch bending coax plants into bushiness, lankiness, or anything betwixt

"Apical dominance" sounds sadomasochistic, but no reason to shudder: it's practiced by plants and, even when carried to an extreme, results in something as agreeable as a head of cabbage. True, we gardeners sometimes have a hand in apical dominance, but it's still just good, clean fun.

Look upon it as hormones gone awry or as hormones doing what they are supposed to do; either way, apical dominance is the result of a hormone, called auxin (AWK-sin), that is produced in the tips of growing shoots or at the high point of stems. Traveling down inside the stem, auxin sets off a chain of reactions that puts the brakes, to some degree, on growth of side shoots, giving the uppermost growing point (the apical point) of any stem the upper hand in growth.

Side shoots mostly arise from buds along a stem, and whether or not a bud grows out into a shoot depends on how close the bud is to the source of auxin; the closer to the source, the greater the inhibition, how far and to what degree depend on the genetics of the plant. 'Mammoth Russian' is a variety of sunflower that grows just as a single stem capped by a large flowering disk; with no side branches at all, this variety demonstrates an extreme example of apical dominance. At the other extreme would be one of the shrubby species of willows that keeps sprouting side branches freely all along their growing shoots.

Even within a single species of plants, individuals vary in their tendency to express apical dominance. Fuchsia varieties such as 'Beacon Delight' and 'Blue Ribbon' express strong apical dominance, so are easier to train upright into miniature trees ("standards") than are trailing varieties such as 'Basket Girl, and 'Blue Satin', which have weak apical dominance. The latter are more at home with their branches sprawling over the edges of hanging baskets. Only with the help of a stake and diligent pruning of side branches can 'Basket Girl' be coaxed to take on the shape of a miniature tree.

There are times when I want to thwart apical dominance: to get a stem to grow side branches, for instance. One way to do this is by removing, if temporarily, the source of auxin—by pruning. The effect is only temporary because a new, uppermost bud soon establishes itself as apically dominant.

For the least possible pruning, I'll just pinch out the soft growing point of a shoot with my thumbnail. This quick and simple check to auxin flow not only causes growth to falter briefly, but also causes lateral buds that were dormant to be awakened into growth. I pinch back the main shoot of coleus plants to make them grow bushier, which happens coincidentally when I harvest basil.

Another reason I pinch out a shoot tip is to slow growth. For example, in late summer I pinch out the tips of my tomato plants to re-direct their enthusiasm for stem growth in late summer into ripening fruit. (By then, tomato plants are growing like weeds, so the effect of the pinch is short-lived, and needs repetition to be effective.) The sprouts of my Brussels sprouts plants swell more quickly in response to my thumbnail's work on the very top of the plant's upright stem. I pinch out the tips of side branches of my young apple tree that turn upward to threaten the dominance of the single leading shoot (the "central leader") that I've selected for the growing young tree.

Whether done by me, you, insects, or diseases, plant response to pinching is relatively quick. Experiments with pea seedlings show that lower buds are stirred into activity in as little as four hours after the apical bud is removed.

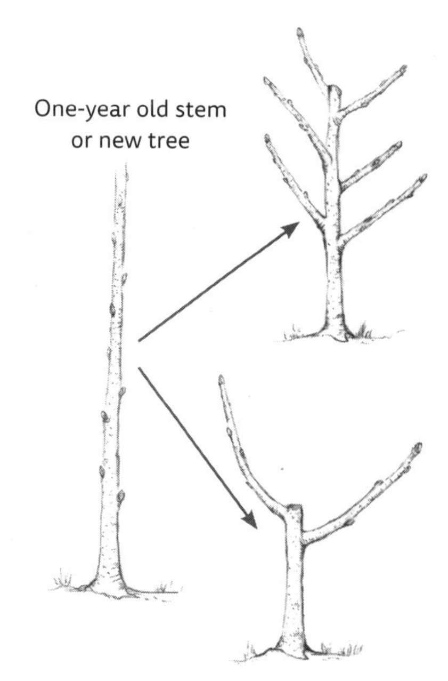

Effect of degree of heading cut on plant response.

Cutting back a larger portion of stem differs from pinching in the degree of response. The more drastically a stem is cut back, the fewer side shoots awaken, but the more vigorously each side shoot grows. This type of cut is called a "heading cut," and plant response depends on the degree of heading. The more inherently vigorous a young stem is before it is headed back, the more vigorous the response to such pruning. Inherent vigor is greater the more vertically-oriented a stem, and the younger it is.

I have a long privet hedge that needs to be sheared every few weeks through summer to keep it to the shape and size I want. Shearing the hedge is, essentially, making hundreds—no, thousands—of slight heading cuts. The plants respond just the way I hope, with dense growth of many short shoots.

At the other extreme is the pruning my butterfly bushes get. Lopping the bush's stems almost back to ground level late each winter is how to coax forth each summer's long, graceful, flowering stems.

On fruit trees, my pruning cuts vary in severity, with the goal of coaxing some—but not too much—new growth on which to bear fruit to replace decrepit older growth. The degree to which I prune depends on the kind of fruit tree. The apple trees, for instance, get a relatively light pruning because they continue to bear well on branches even a decade old. I go at my peach tree more aggressively because peach trees bear only on one-year-old stems, so each year they need enough new stems coaxed for a good crop the following year. The bearing habits and the pruning of plums lie somewhere between these two extremes.

I've used pruning (i.e., harvest) to get multiple harvests of cabbage from a single plant. A cabbage head is actually a stem that has been foreshortened, with side buds and their attendant leaves close together, one above the next. So lopping off the head during harvest is, essentially, a heading cut. Harvesting releases side buds from apical dominance so that some of them can then sprout to make new heads. I leave cut cabbage stumps in place after harvest, then let three or four side sprouts grow, snapping off the rest to strike a reasonable balance between the number of new heads and their eventual size.

Many a homeowner deals with an overgrown tree—one that has grown to block a window, for example—by irreverently hacking back the tops of branches that are blocking that window. Said homeowner then bemoans the dense and vigorous sprouts that regrow and quickly block that window again. More effective would be "thinning cuts," which are cuts that totally remove offending branches right to their origin. With no buds left to regrow, apical dominance is then moot. Plant response is: nothing, nada, zip—near the cut, at least.

I use thinning cuts for congestion (on my plants). When too many stems clog up the interior of a tree or shrub, removing some of them right to their beginnings opens up the space without eliciting regrowth at the cuts. For that aforementioned tree blocking a window: not only do thinning cuts not result in regrowth, but they also can be conveniently made closer to ground level, at the origins of the offending stems.

A downside to either heading or thinning cuts is that it stunts plants, to some degree. Stems contain some stored energy and, if it's a time of year when the stem is clothed with leaves, pruning also lops away some of the plant's energy-producing "factory." Of course, one reason to prune might be to keep a plant small. And pruning is often a necessity, even if I'm not trying to keep a plant small.

Because auxin is produced at the highest point of a stem, an elegant way to thwart or re-direct apical dominance without having to prune off any part of a shoot is to bend it down. If the dampened stem is bent in an arch, a bud at or near the high point of the arch becomes top dog and sprouts to become, usually, the most vigorous shoot. On the other hand, if the tip of a stem is lowered while being kept straight, that is, without creating an arch, then each bud along the stem is only incrementally higher than the one before it. In this case, the uppermost bud expresses weak apical dominance and many buds lower down along the stem are released from inhibition to make relatively weak growth. The more vertically oriented the shoot, the fewer buds awaken and the more growth from each bud, especially those nearer the tip of the stem. The more horizontal the branch, the more buds awaken and the less growth from each. So I can regulate the amount of growth and side-branching with appropriate adjustment of branch angle, changing it, if necessary, as I watch my plant grow.

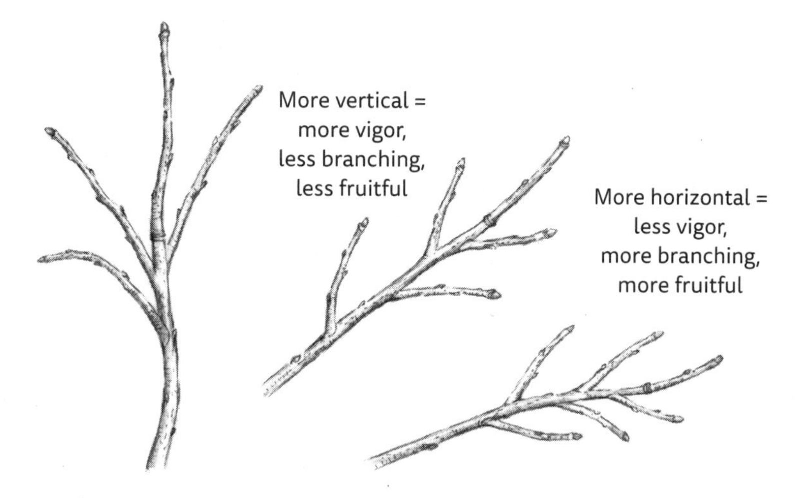

Branch orientation and apical dominance.

Branch bending is used to the extreme to create "en arcure" espaliers, in which all stems are bent over. The single, upright stem of a new tree is bent in an arch. The vigorous stem sprouting from the high point of that arch is likewise bent, in the opposite direction, and so on, the arches alternating their directions as the

Pruning Fruit Trees

Pruning and branch bending are especially useful with fruit trees during their training years and then when they begin to bear fruit.

Apical dominance comes prominently into play when training a fruit tree to a "central leader" form, which is much like a Christmas tree. A single stem, the "leader," grows upright, with side branches of decreasing length with increasing height up along the single central leader.

Shortening the leader of a young fruit tree is the way to encourage growth of side branches that become future limbs. A young fruit tree, whether I propagate it myself or purchase it, is often nothing more than a "whip" whose tree-ness is nothing more than a single, vertical stem: the trunk-to-be. A heading cut that removes a quarter to a third of a young whip's length produces the desired three to four side branches of moderate vigor. The bud just below my pruning cut typically grows most upright and vigorously, continuing skyward as an elongation of the leader, which will be shortened back again a year later to encourage more side branches.

One caution: fruits forming high on a young, developing central leader could weigh it down, bending it over, and thwarting its apical dominance. Then a new stem might attempt to take that role, which makes for a poorly trained tree. I re-establish the original leader's leadership role by either removing or reducing the number of offending fruits, or by staking the stem upright.

A tension exists between shoot growth and fruiting in plants, with more of one associated with less of the other. Bringing a side branch to or near a horizontal position while keeping it straight not only awakens growth in many buds along that stem; the resulting weak growth has a tendency to develop into flower and fruit buds rather than buds that would grow into shoots. Bending side branches is one way to coax apple trees and pear trees into earlier production.

Espalier en arcure.

young tree grows. A row of dwarf apple trees might be woven together as each successive tier of bent branches is tied to arches on neighboring trees on alternating sides.

Wherein I make designs with the traceries of my fruit plants' branches

An espalier is a plant whose branches are trained to an orderly and ornamental, usually two-dimensional form. Hang fruit on those branches and the result is a plant offering beauty and good eating. Pruning and branch bending allow each branch to be

Pear espalier in northern France.

Training and early pruning of T-shaped red currant espalier.

bathed in abundant sunlight and air, resulting in fruits that are large, luscious, and fully colored. To me, an espalier represents a most happy commingling of art and science.

Espaliers have a long tradition in northwestern Europe, especially with apples and pears. In regions where the day length and climate are quite different, such as over much of continental North America, attempts at espalier often fall short of expectations. Sure, branches can be bent in all sorts of designs, but they won't necessarily be well-clothed in the requisite flowers and then fruits. (Two exceptions: Asian pears are so eager to bear fruit that they work as fruitful espaliers everywhere, and the same can be said for some of the newer, "spurry" varieties of apples.)

One fruit plant that works especially well as an espalier everywhere is red currant. Red currant espaliers also need to be pruned

Maintenance pruning of T-shaped red currant espalier.

only twice a year, in contrast to the monthly or more frequent pruning sessions demanded by apple and pear espaliers. Adding much to the show, red currants' bright red fruits dangle from the branches like translucent, red jewels. The espalier technique that I am about to describe can also be applied to white currants and to gooseberries.

My red currant espaliers decorate the fence enclosing my vegetable garden. Each plant is trained to the shape of a T: a single trunk capped by two fruiting arms that grow in opposite directions along the fence, which provides support.

For the trunk, I chose the strongest shoot on my plant, removed all others, and then tied that retained shoot to the fence to keep it upright and vigorous. That tying upright of the trunk-to-be keeps that shoot apically dominant, which is something shoots on shrubs, such as red currants, don't express a lot of—that's why they're multi-stemmed (shrubs) rather than single-trunked (trees). Any side shoots that attempt to grow are either lopped off or temporarily pinched back before being lopped off. Side shoots speed thickening up of the trunk.

Once this trunk-to-be grew just above the top of the three foot high fence, I cut it back to the three-foot height, in so doing releasing the remaining buds from the suppressing effect of apical dominance. I selected two shoots that started to grow from the upper portion of the trunk to become fruiting arms, training them along the fence in opposite directions as they grew, and removing all others. To keep these developing arms growing vigorously, I left their ends free to turn upward as I tied portions closest to the trunk to a horizontal position. Alternatively, the growing arms could each be lashed to a length of bamboo to direct the growing arm upward and outward from the T at about a 45° angle. As the arms lengthened and less extension growth was wanted, the bamboo, with the attached arms, could have been lowered until they were horizontal.

Maintenance pruning and fruiting began even as arms were developing. The arms, because of their horizontal orientation, exhibit little apical dominance so side branches developed freely along their lengths. Two simple cuts keep the form neat while encouraging abundant fruit production. First, just before the berries start reddening, I cut back all those side branches to about five inches in length. I perform the second cut in winter, cutting those side branches back again, this time to about an inch.

The only problem with my red currant espaliers, which hang onto their berried treasures for weeks, is that picking the fruits steals from the plants' appearance.

Questioning the advice to put the brakes on tree growth with summer pruning

"Prune when the knife is sharp" goes an old saw. Not true. (But it's never a good idea to prune if the knife is not sharp.) How a plant responds to pruning depends not only on how much is cut off, but also on when the operation is done. The usual advice about pruning woody trees, shrubs, and vines, including that offered above, concerns pruning them in "winter," which is anytime that they are dormant and, in the case of those that are deciduous plants, leafless. How about pruning at other times?

Sometimes, trees, shrubs, and vines grow very well, too well in fact, so that their stems are crying out to be pruned. Enter summer pruning, often recommended as a better way than winter pruning to quell over-exuberant growth.

In the heat of the latter part of summer, woody plants should be getting ready to prepare for winter. Peering at some shoots of, say, a crabapple after midsummer, I see that rather than unfolding new leaves at their tips, their terminal buds just sit there, fattening up. The plant's energies have been directed to making shoots thicker and more woody, not longer. Food energy is being stored up in their stems, trunks, and roots, energy that will be needed the following season to fuel early growth of new shoots and leaves, until new leaves are mature enough to not only feed themselves but also to export food to other parts of the plant.

Although summer pruning generally is not—not!—a good idea, it does have a place in gardening. The deep red color of a ripe apple (a variety that ripens to red) needs a direct hit of sunlight; removing some stems in summer to allow fruits to bathe in increased sunlight results in prettier apples. Don't count on sunlight to paint the ripe color on every kind of fruit, though; plums, cherries, and grapes color up when ripe whether or not the fruits themselves bathe in light.

Removing a few stems here and there on any plant in summer also lets air in among remaining stems, leaves, and fruits, helping them to dry more quickly following rain or dew and so lessening the threat of disease. Even clipping off just a few leaves, sometimes recommended near clusters of grapes, can be beneficial in this way.

But all this summer pruning is not absolutely necessary. More judicious dormant pruning, such as cutting away enough branches on an apple tree to let the sun shine in on all the remaining ones, might obviate the need for summer pruning.

Some styles of growing plants necessitate summer pruning by their very nature. Hedges, for example. My privet or yew hedges can't keep their shapes without being pruned repeatedly from early spring right through summer. Espaliers need summer pruning so that the tracery of their stems remains decoratively prominent and each of those geometrically trained branches is thoroughly clothed in fruits as well as in leaves.

And yes, there are times when summer pruning is needed for no other reason than to contain overly vigorous stem growth. Dormant, woody plants respond to heading cuts just as soon as they begin growing in spring. As the growing season marches on, though, with plants shutting down in preparation for winter, their buds are less inclined to sprout in response to heading cuts. This nonresponse has given rise to the traditional belief that summer pruning is more dwarfing than dormant pruning. Ah, but what about the following spring? That's when the summer-pruned shoot decides to respond. Plants have an amazing capacity to act however they please no matter what we do to them.

Okay, summer pruning sometimes can be more dwarfing than dormant pruning. Regrowth following summer pruning earlier in the growing season can be pruned again and again, removing food-producing leaves to sap a plant's energy and dwarf it.

And under certain conditions, summer pruning also can prompt the formation of flower buds rather than new shoots. That's the idea behind creating fruitful espaliers. Except, as I wrote in the previous section, that summer pruning might work in one region to set up fruit buds but might not do so in another, where climate, day length, and/or other environmental factors differ. Depending on the summer weather that has preceded and followed my summer pruning, responses of apple trees to pruning techniques from northern Europe that I have attempted to espalier have ranged, right below the cuts, from some die-back to vigorous (unfruitful) growth to, occasionally, the desired swelling of fruit buds.

Response to summer pruning can vary depending on when, during the growing season, a stem is pruned. Not so for dormant pruning; response is relatively consistent no matter when, during this period, a stem is pruned.

I also consider plant health when deciding when to prune. The weakening effect of repeated summer pruning could do in an already weakened tree. My peach tree gets pruned early in the growing season, when just in flower, or right after flowering. Peaches are particularly susceptible to infections at open wounds; early season pruning exposes a stem wound for the minimum amount of time before healing begins. Wounds in summer also heal quickly. The best time to prune a diseased or damaged branch is anytime it's noticed.

On the genesis, reason for, and propagation of weeping trees...

Why are these trees so sad, even with pink or white blossoms cheering up their branches? But, of course, they're not really sad. They're just weeping.

So why are these trees weeping, then, even if they are not sad? They weep because they grow downwards. Instead of young stems reaching for the sky, as is the case with most trees, young stems of weeping trees toy only briefly with skyward growth before arching gracefully down towards the earth. Some plants begin to weep in earnest only after they get some age to them.

A weeping tree may have begun life as a chance seedling whose quirky arrangement of genes directed its stems to weep. Some such plants, although woody, could hardly be called trees. A weeping kind of goat willow, for example, makes a billowing groundcover.

Or, a weeping tree may have begun life with normal stature—until some cells in a branch of that tree underwent a slight mutation to a weeping habit. Perhaps the mutation was due to the effect of sunlight or temperature, perhaps the mutation was spontaneous. At any rate, all new stems and branches originating from those changed cells weep, in which case only part of the tree would be weeping.

In either case, a number of genetic quirks could account for various kinds of weepiness. Generally in plants, a stem pushes skyward in response to gravity and light, gravity being the stronger influence. In some weeping plants, the stems grow actively downward, following gravity. Depending on the mutation, these plants' roots might actually grow upward—not a healthy condition for a plant! Another kind of mutation might more correctly be termed creeping in that the stems are oblivious to gravity. They might eventually turn downward, but only from their own weight. So-called "lazy" mutants weep just because they don't have the wherewithal and then the strength to point upward.

So how can I tell if my plant is weeping, creeping, lazy, or something else? One way to tell if it's weeping would be to turn the (potted) plant on its side or upside down; a weeper's stems will actively grow in the direction of gravity.

Botanists don't know all the mechanisms that cause plants to respond positively or negatively to gravity. Statoliths, which are starch-filled granules within cells, have been implicated as one sensor. Their heaviness causes them to sink to the lower side of cells and then transmit a signal for increased cell elongation on one side or other of the stem or root tip. The hormone auxin comes into play, in this case inhibiting cell elongation on the lower side of the root tip, so the root grows downward. Auxin has the opposite effect in stems, resulting in upward growth. Once cells are vertically oriented, statoliths are in their normal position and not favoring elongation on one side or the other of a growing root tip or stem. "Lazy" weepers lack sufficiently dense statoliths. (Much of this information comes by way of research on travel in outer space.)

Let's go back to using the name weeping for any plant whose stems are weeping, creeping, lazy, or otherwise staring at the ground. How can I make whole new plants from that weeping seedling or those weeping branches? If the weeping plant is one that roots easily from cuttings, I could just clip off a branch, stick it in the ground or some potting soil, and nurture it along.

But could I really appreciate the weepiness of a plant if it keeps pressing downward? Not necessarily. So I might want to tie a stem to a stake to help it get some distance off the ground to eventually better express its weepiness.

Another way to propagate a weeping plant—or one that does not root easily from cuttings—is by grafting a stem or bud from the weeping plant atop a trunk of some upright plant. As with any graft, success is possible only if the trunk section is closely related, botanically, to the weeping stem piece. A friend's weeping cherry tree, for example, was created by grafting a stem from a weeping cherry atop a five foot length of trunk of some upright cherry. The height at which the plant begins to weep was determined by the height at which it was grafted. The graft juncture remains at the same height and is usually obvious throughout the life of the tree.

Sometimes a branch of a weeping tree will all of a sudden start reaching skyward. (Talk about a wacky looking tree!) This situation has three possible explanations and one cure. That stem could have originated from a bud on the non-weeping rootstock just below the graft. Or, a weeping tree that originated as a branch mutation might have retained some non-weeping cells that occasionally express themselves in upright branches. Finally, it's possible for a weeping branch to mutate again, this time reverting to a non-weeping form. In any of these cases, cutting back any non-weeping stem right back to its origin makes the tree pretty again.

Even looking beyond the ubiquitous weeping cherries and crabapples, weeping trees are not all that rare. Japanese dogwood is a lovely tree whose white blossoms unfold after the leaves are fully out; the variety 'Elizabeth' has somewhat weepy upper branches. A weeping form of katsura tree presents a waterfall of bluish green leaves. Weeping forms of European beech, white ash, birch, and hawthorn have been selected, named, and propagated, as well as weeping evergreens such as white spruce and boxwood. Few weeping evergreens are more graceful than the 'Sargent' hemlock. Or more odd than a weeping form of giant sequoia, whose leading stem pushes skyward in fits and starts, zigging and zagging and dipping along the way but always remaining clothed in a shaggy mane of droopy branches.

Weeping Atlas cedar tree.

Many landscapes benefit from some weeping tree, whether it's a willow along a stream bank, a weeping cherry lending grace and tranquility to a front lawn, or a weeping beech providing a hideaway for kids. The only caution with weeping trees is not to plant too many, which might be no more than one. Otherwise, the scene can look sad indeed.

A comfortable seat in a sunny spot gets trees and shrubs ready for winter...

A pleasant activity—especially on a warmish, bright sunny autumn day—is getting trees and shrubs ready for winter. Here is what I do: move a chair to a sunny spot, get a cup of hot tea, then sit down and relax. Getting trees and shrubs ready for winter really is mostly about not doing anything.

For example, reiterating what I wrote previously about pruning late in the growing season: I don't.

I also avoid fertilization. Fertilizer, like pruning, can stimulate growth, in this case of the whole plant. Again, what is needed instead is for plants to be shutting down for winter.

And finally, in autumn, I don't water, or at least am careful about watering. Especially after a long period of dry weather, during which plants lapse into a semi-dormant state, water gets plants growing again, which is not good this time of year.

I strike a balance with watering, though. I don't want active growth, but I also don't want any plants going into winter thirsty. In many regions, autumn rains can usually be counted upon to carry plants into winter. One exception would be newly planted trees and shrubs, whose roots cannot yet reach for sufficient water on their own.

Okay, sometimes I really want to get out of that sun-basked chair and get my blood moving. Perhaps I want to try out my new pruning shears? I could go ahead and prune any plants that are super-hardy, such as rugosa rose, gooseberry, honeysuckle, and witch hazel. And also prune, if I wish, plants such as St. John's wort, vitex, and Russian sage, which naturally die back or will be

A tree protected for winter.

cut way back anyway before spring. And I would have to do some pruning on hybrid tea roses and any other cold-tender plants that need protection in order to fit them into whatever swaddling they need to protect their stems in winter. Any of this latter pruning is best delayed until the weather turns cold, so that plants get some exposure and acclimatization to cold.

Despite the above admonition against fertilization in autumn, if I really want something to do, I could get a jump on spring and spread fertilizer in autumn—but only if I choose the fertilizer carefully. What is needed is a fertilizer that plants cannot absorb until spring. Most organic fertilizers fill the bill. The nutrients in an organic fertilizer such as soybean meal, for example, remain locked up until released with the help of microorganisms in warm, moist—that is, spring —weather.

And then if I need more to do outdoors, I can paint the trunks of young trees. A coat of diluted white, latex paint reflects sunlight to prevent winter sunscald of thin bark. Mice or rabbits are always threats, so I protect the lower trunks of trees with cylinders of one-quarter inch mesh "hardware cloth." Or I use purchased tree wraps or other types of trunk protectors that do both jobs, keeping sun and rodents off the trunks. I've even painted the trunks with eggs added to the latex paint mix. The eggs make the bark unattractive to mice and rabbits, which are vegetarians. Sometimes I throw in some cinnamon, mint, and/or rosemary for added aroma that, I hope, makes the mix even more repellent to the furry creatures.

Enough, enough. Time to go back and sit in the sun with my tea, letting my plants ready themselves for winter.

In which it is demonstrated that buds are not boring

Winter is a good time to look at some of the finer details of trees and shrubs—their buds, for example. Buds!? Bo-r-r-r-ring, you say? Not really, if you take the time to appreciate details such as their shapes, colors, and textures.

Buds can do more than just help wile away winter hours. They can disclose a plant's identity as well as foretell what the upcoming growing season holds in flowers and fruits, even provide some guidance for propagation of new plants and for pruning.

Locked within every bud are the beginnings of a shoot or a flower, partially developed and telescoped down into a compact package. Except for adventitious buds, more common in some plants than others and popping out haphazardly on stems or trunks, buds appear along stems at distinct locations, called nodes. A node is where a leaf was attached last summer, and this winter's buds were formed in the crotch where last summer's leaf stalks and stems joined. In some plants, including fragrant sumac and black locust, the node is apparent from the leaf scar even though the bud itself, buried within the stem, is not visible in winter.

Noting the arrangement of buds along a stem is the first step to identifying a leafless tree or shrub in winter. Buds on some plants arise directly opposite each other. On other plant, buds are "alternate," that is, arising singly on alternating sides of the stem.

Because buds can grow out into stems, that opposite or alternate bud arrangement is mirrored in a plant's stem or leaf arrangement. Not always, though, because not all buds awaken into stems in spring; some remain dormant and others open to become flowers. And then there are stems that die and fall off. Plants with adventitious buds can have stems popping out just about anywhere.

It turns out that most deciduous trees have alternate buds. So if I happen upon a leafless tree in winter with opposite buds,

I can pretty much bank on its being some species of maple, ash, dogwood, or horsechestnut (which some people find easiest to remember with the acronym MAD Horse). Of course, once I identify a tree as, for example, a maple, I have to look for other details, such as the bark, to tell if it is a red, sugar, silver, or Norway maple.

(A few less common trees also have opposite leaves, including katsura, some species of eucalyptus when they are young, paper mulberry, lyontree, devilwood, cork tree, olive, catalpa, fringetree, and paulownia. Most shrubs have opposite leaves. Some of the aforementioned trees, such as fringe tree and devilwood, with opposite leaves, straddle the fence in stature and growth habit between that of trees and shrubs.)

Both for winter entertainment and for practical reasons, I'll take a closer look at the buds of some trees and shrubs. They vary in color, size, shape, and texture: witness the elongated, mahogany buds of pussy willow, the brown velour buds of pawpaw. Some plants—viburnums, for example—have naked buds, enveloped only by the first pair of leaves, rather than the scaly covering protecting the buds of most other plants.

Mature plants have two kinds of buds. Come spring, those that are longer and thinner will expand into shoots. Flower buds are usually fatter and rounder. I note how dogwood flower buds stand proud of the stems like buttons atop stalks. I take a look at a peach branch with its compound bud: a single, slim stem bud in escort between two fat flower buds. Apple and crabapple flower buds occur mostly at the ends of stubby stems, called spurs, that elongate only a half an inch or so yearly. Pears likewise bear flowers and fruits on spurs. Spurs have to be old enough and their buds fat enough before flower buds are contained within, which always makes it frustratingly hard for me to predict the pear crop for the season to come.

Because many plants lay down the beginnings for any year's buds during the previous year, even in winter I can tell what kind of flower show or fruit crop to expect, barring interference from late frosts, insects, diseases, birds, or squirrels. And if what was billed by the plant to be a good year turns out paltry, I can lay blame on spring weather or some pest. Looking at buds also is one way that I can assess how effectively I pruned the previous winter.

Pruning stems laden with flower buds is one way I reduce—on purpose!—the fruit crop for the coming season on my peach tree so that more of the tree's energy can be directed to making the remaining fruit larger and more luscious.

For propagation, stems used for cuttings are more likely to root and scions used for grafting are more likely to "take" if they lack flower buds or, at the very least, have some shoot buds. After all, the goal is for these cuttings to grow rather than fruit, at least for a season or so.

How buds become burls and witches' brooms

As I glance upward into tree limbs in winter, perhaps searching for some further way of identifying a tree or sign of spring in a swelling flower bud, my eyes are sometimes arrested at a fat, rounded growth on the bark. On some trees, these hard, woody outgrowths—called burls—stand out on an otherwise clear trunk like a goiter. On other trees, the whole trunk might be covered with these masses.

Burls are no cause for alarm. Little or no harm befalls the tree.

That said, burls might—just might—indicate that the tree has been under stress. All sorts of things have been implicated in the occurrence of burls. For instance, a burl might grow in response to a limb once rubbing the bark, insect chewing, or some other physical injury. Perhaps the tree experienced or is experiencing some environmental stress—temperatures too cold or too hot, a nutrient deficiency, or not enough sunlight, for example.

Burl on tree bark.

Diseases have also been implicated in the formation of burls. However, no pathogens are found inside burls when they are cut open. Still, a pathogen could have done its job of inducing a burl, then skipped on to other adventures.

Genetics might also be to blame, because some tree species are more prone to developing burls than others. Redwoods are renowned for their burls, which are often sold as souvenirs.

All these "coulds," "mights," and "been implicateds" are telling: no one really knows for sure just what makes a tree make a burl.

We do know, though, how burls develop. Along any stem are buds that can elongate to become shoots, and each of these shoots similarly has buds that can become shoots themselves, such growth proceeding *ad infinitum*. Obviously, not all buds on a plant stretch out; some remain dormant, for the time being at least. In the case of a burl, instead of dormant buds expanding into straight shoots, they grow inward, twisting and turning under the bark, and never emerging as branches.

On the rare occasion when a shoot does elongate from a bud within a burl on a mature tree, the shoot usually expires anyway from lack of sufficient light. Interestingly, redwood burls frequently sprout when cut from the tree and placed in a pan of water, as if the water reminds the cells that they can become real, elongating shoots. The miniature redwood forest that develops can grow on to become a redwood tree if the burl is transferred to soil and the number of sprouts is reduced. On a tree, though, cells within any burl generally just keep dividing with no obvious purpose or benefit to the tree.

Benefit to us humans is another story, though. Cut open a burl and instead of straight grain you find waves and swirls of wood, marbled and feathered wood, perhaps some "eyes" staring back at you. Woodworkers lust for burls, whose swirling tissue, no longer constrained within the straight and sharp-edged geometry of sawn lumber, can be turned on a lathe into highly figured bowls. Burls are also sliced into veneer (or shaped and sanded smooth for the rounded newel caps I made for the stairway in my home).

The beauty within a burl and the theories for their cause could prompt one to try to induce them on a tree. Yet no one has been able to do so reliably. So burls are harvested from trees on which they have naturally formed. Because harvesting a burl can leave a large scar, cut burls only from trees that are being or have been cut down for other purposes.

Looking higher up into a tree—for maximum effect when the tree is silhouetted against the night sky on Halloween—I might see something spooky: a witches' broom. I'm talking about a real witches' broom. No witch, though.

Witches' broom is the name given to broom-like growth that sometimes occurs on part of a plant. How does this broom form? Branches growing off a normal stem typically originate a few inches apart along the stem. Now, if that stem is telescope down so that only a fraction of an inch lies from one branch to the next...*voila*, what's left is a myriad of branches originating in close proximity—a broom!

Witches' brooms usually result from infections by certain viruses, fungi, and parasitic plants (such as mistletoe). The infection probably causes the broom-like growth by upsetting the balance of plant hormones in the stem. No longer do the apical buds of stems suppress bud and stem growth nearby.

Another hormone, gibberellin, also might be going haywire in a witches' broom. This hormone causes cells to expand once they have formed. The compressed stem of a witches' broom may indicate insufficient gibberellin. The cells form but the stem never stretches out. (The opposite occurs with what is known in Japan as *bakanae*, or "foolish seedling" disease, of rice. Plants infected by the fungus *Gibberella fujikuroi* "foolishly" stand above the rest in a field, growing so gangly that they eventually topple over. The cause, in this case, is too much gibberellin.)

Some witches' brooms harbor an infection that eventually kills the plant. Other strange things also can happen, such as when potato plants with witches' broom form tubers above ground. Pretty spooky, eh?

Many witches' brooms do no apparent harm to a plant, and have even given rise to attractive, or at least interesting, garden plants. If a branch from a witches' broom is cut off and either grafted on a compatible rootstock or rooted, the resulting plant retains the compact growth habit of the broom. Occasionally, witches' brooms are spied high up in tall trees. These sky-scraping brooms are sometimes "bagged" for propagation with the well-placed shot of a rifle, then gathered after tumbling down to terra firma.

The black pine variety 'Hornibrookiana', which after 30 years will be no larger than six feet across and two feet tall is a witches' broom, as is the Norway spruce variety 'Maxwellii', which also remains a somewhat flattened cushion of green, this one maturing at a few feet tall. And 'Beuvronensis' Scotch pine is a dwarf globe of green that never grows larger than 4 feet across and 4 feet tall. All these plants remain dense and green, and never grow out-of-bounds.

So, you see Dorothy, just as there are good witches and bad witches, so there are good witches' brooms and bad witches' brooms.

Dwarf spruce tree from witches' broom.

On entreating and helping trees to stay asleep

Each warm spell of late winter makes the next blast of cold feel that much more frigid to me and—more importantly—to my trees and shrubs. Once winter sets in, these plants would be much better off if the weather would stay cold until it was ready to turn warm in earnest. Except in some alpine and very cold latitudes, low temperatures in late winter are usually the only thing keeping trees and shrubs secure in their winter sleep.

Such is not the case in autumn. Short days and cooling temperatures put these plants into a kind of sleep from which they can't awaken until they feel that winter has passed. The hormone, abscisic acid, builds up in autumn and is largely responsible for lolling plants to sleep. (The effect of abscisic acid, discovered in 1963, is often opposite to that of auxin. Abscisic acid keeps seeds dormant, closes stomates during periods of water stress, and arrests shoot growth, but not root growth, which it may even stimulate.)

Plants mark time in winter by the amount of cold they experience, each kind of plant requiring different amounts of cold before it can awaken. During cool weather, abscisic acid breaks down slowly in plants, slowest in plants with longest chilling requirements. So even with warm temperatures, for example, dormant apple trees of most varieties will not start growing again until they've experienced an accumulated total of about 1,000 hours of temperatures between 30 and 45 degrees Fahrenheit.

Plants native to regions where winters are mild enough not to harm new growth, should it occur, need very little chilling before they can resume growth in spring. 'Beverly Hills' apple, suitable for warmer climates, resumes growth after only 300 hours of chilling.

Little chilling is also needed by plants native to regions where winters are long and steadily frigid. In such climates, chilling requirements are not fulfilled until late spring. Then, the plants must begin growth quickly in order to ripen their seeds within the short growing season.

In regions with a continental climate, such as much of the US, autumn weather puts some hours into fruit trees' chilling "banks" before winter is even under way. With typical fluctuating winter temperatures, "deposits" continue slowly through the winter. Once that "cold bank" has been filled, warmth is free to awaken these plants, which is exactly what I fear happening each spring during early warm spells towards winter's end. In this kind of climate, "low-chill" plants are especially apt to awaken too early in spring. New growth, being succulent and lush, is very susceptible to cold. Frost biting at new stems weaken a plant, could even kill it, depending on the vigor of the plant and the timing and degree of cold.

Early flowers are part of that awakening growth. If nothing more, cold threatens to turn colorful petals to brown mush. Worse, dead flowers can't go on to become fruit, leading to further disappointment if the hope was for something again decorative and, eventually, luscious. Hence, apricots, native to the colder regions of Manchuria, are often an iffy crop to grow where chilling requirements are partially or wholly fulfilled in autumn and during cool, but not frigid, spells of winter—their eager blossoms are often killed by late spring frosts.

No need to sit back and do nothing about the weather. It is possible to help plants be less susceptible to the ill effects of early warm spells.

The easiest approach is to grow plants adapted to the region. Even within a species, individual plants vary in their winter cold requirement. A red maple native to Georgia, for example, is more likely to awaken prematurely in New York than would a red maple that is native to New York. New York red maples having "early wakening" genes would have died off.

Growing only adapted plants is a tough order for any diehard gardener. I want to grow figs, peaches, apricots, southern magnolia, camellias, and other plants not really adapted here. Most would just be killed by winter cold. The problem here with peaches and apricots, however, is that they wake up too early; my goal is to try to keep such plants asleep as long as possible. So I look around my yard and note how snow lingers longest on the north side of my house or on north-facing slopes or on the north sides of buildings. Growth will begin later in spring there. Even a tree or shrub that needs sun can thrive in such situations because such sites become increasingly bathed in sunlight as the season advances, with the sun rising higher in the sky and wrapping further around the horizon until the summer solstice.

One winter that brought too many warm spells sent me outdoors with a sprayer filled with a suspension of kaolin clay (Surround®, which also controls certain pests) to apply to susceptible plants to keep them asleep longer. Diluted, white latex (not oil!) paint would have worked equally well. The white color either spray puts on stems and buds reflects sunlight to keep everything cooler.

Water resists temperature changes, so proximity to large bodies of water or to air that's travelled over large bodies of water keeps things cooler in late winter and early spring. That's why gardeners in coastal regions and in western Europe, the latter's climate moderated by the Gulf Stream, have fewer problems with overly eager plants in spring. Before I consider packing up plants for a move to coastal Ireland, I keep in mind that water masses also subdue summer heat, which makes ripening heat-lovers like figs, peaches, and tomatoes to perfection more of a challenge.

About a quick and easy way to hasten spring

Anxious nail-biting isn't my only association with chilling. Once a plant's chilling "bank account" has been filled, the readiness of its stems to flower can be put to more convivial use. All that's needed is warmth, so, brought indoors, flowering stems are urged to unfold their buds for a colorful show of flowers—an early spring, in a vase, on my dining table or window sill.

Starting with branches of flowering trees...even though leafless in winter, some trees and shrubs do notice when days are getting longer, their sleeping buds jerking awake as soon as they feel day length has become sufficiently long. Trees most affected by photoperiod—that is, day length—include birches, some species of maple, sycamore and London planetree, black locust, elm, and American beech. Least affected trees include ashes, gingko, pines, pin oak, and, in contrast to the American species, European beech. White oak, red oak, linden, and ironwood are among trees intermediate in response. Plant species, as well as duration and amount of cold, all interact with photoperiod in its influence on awakening plants in spring. (Photoperiod can also be a stimulus for plants to enter dormancy in autumn, a response that can affect winter hardiness of trees exposed to artificial light, such as street lights.)

Apples, plums, and most other fruit trees—all of whose branches bear beautiful, sometimes fragrant, blossoms that can be forced indoors—wait only for temperature changes to tell them when spring has arrived.

Branches snipped from trees and shrubs and brought indoors for forcing can be exposed to more steadily warm temperatures than the fluctuating warmth of late winter and early spring outdoors. I just have to wait long enough to begin forcing blossoms from these branches.

Even then, to force any branch well demands patience. Rush the process, and blossoms open sporadically along the branch, then dry up and fall off.

To stay alive and flower, a cut stem needs water, which is no longer afforded by roots. To prevent drying, I first plump up all the buds by completely immersing the cut branches in tepid water for a few hours, then recut their bases to expose fresh xylem cells (through which water and nutrients flow), and put them in a vase of water in a cool room. The vase gets moved out for display just as the buds are about to burst into bloom. The time for this will be shorter the closer to the natural bloom time that I begin forcing.

Am I occasionally too impatient to wait for the chilling "deposits" to be fulfilled to even begin forcing? Banking rules for the chilling bank are not all that strict, and a dormant branch might be awakened from its sleepy state with a high temperature shock. Immersing branches for a few hours in 90 degree water usually does the trick. On the living plant, chilling needs can also be reduced by exposing a plant to long days, but this is not as convenient as the warm water treatment.

Hardy, spring-flowering bulbs, such as tulips, daffodils, crocuses, and hyacinths, can similarly bring spring indoors early. Bulbs are not really out of place when considering forcing stems to flower. Most bulbs are, after all, some kind of modified stem.

So forcing flowers out of season from a spring-flowering bulb is not very different from forcing a forsythia stem, the main difference being that a bulb, at least when purchased, lacks roots. But bulbs are fleshier and moister than branches, so can pump some moisture up into a developing flower, and roots can even be coaxed from bulbs in preparation for their show. The first step in forcing a bulb is to plant it, usually in a pot of soil or stones, or to suspend the bulb above water with only its base dipping in. A bulb's roots, like the roots of other plants, grow whenever soil temperatures are above 40°F., so they can be in place and ready to support leaves and flowers when spring comes.

A bulb's flower buds are initiated the previous growing season, just like the flower buds of spring-flowering trees and shrubs. And, like the flower buds of spring-flowering trees, a bulb's flower bud has a chilling requirement and will remain dormant until that requirement is met. So after planting a bulb for forcing, it needs to be kept cool, in that magical 30° to 45° Fahrenheit range.

In time, typically 6 to 8 weeks, which varies with the type and variety of bulb, roots will grow and eventually the flower's winter chilling requirement will be satisfied. Once the chilling "bank" has been filled, bud growth simply awaits temperatures warm enough to grow. A well-grown, spring-flowering bulb comes packed with a flowerbud-in-waiting...waiting, that is, for a chilling period to break its dormancy and then sufficient warmth to allow growth. Keeping a potted bulb cool at this point is useful for staggering flowering for multiple pots of bulbs or delaying flowering for a specific date, such as someone's birthday.

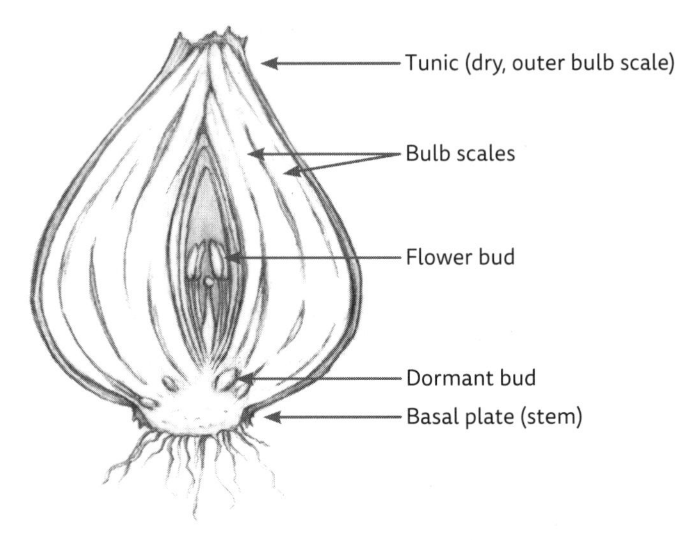

Cross-section of flower bulb.

When I'm ready to enjoy the flowers, I don't just bring the pot into a hot room. The flowers would blast open and collapse. The plants, at this point, need gradually increasing warmth, and enough light to draw out a sturdy flower stalk. Forcing bulbs to blossom out of season demands a certain amount of artistry in addition to science.

The whole forcing process can be bypassed when forcing Paperwhite narcissi. These bulbs hail from perennially warm climes and will bloom without any prior chilling. All that's needed is to pot them up and wait as long as it takes for the fragrant, white blossoms to unfold. To stagger their blooms, pot them up sequentially; lack of water keeps them dormant.

Sunlight is important but sometimes shade offers improvement

Non-woody plants also have stems and leaves—and, in the case of vegetables, we eat them. And some of my vegetables blanch at what I do to them. They don't blanch from fear, but from lack of light. I'm happy they do, and they are none the worse for it. The reason I blanch these vegetables is to make them more tender and, with vegetables having overly strong flavors, to mellow out that flavor.

(Blanching by excluding light is not to be confused with blanching in cooking, which is the brief scalding of, say, a vegetable in boiling water or steam to arrest enzyme action before freezing it for storage.)

Blood draining from our faces—from fright, perhaps—is what makes you and me blanch; chlorophyll loss is what makes plants blanch. Chlorophyll, which gives plants their green color, is in constant flux in plants and light is necessary for its continued synthesis. Chlorophyll and hemoglobin, incidentally, are only slightly different chemically. Hmmm.

Lack of light isn't the only thing that makes plants blanch. If leaves can't get their fill of iron, they show it by turning yellow. (We humans also turn pallid from lack of iron.) Lack of the essential nutrient magnesium also causes blanching, as does sulfur dioxide, an air pollutant, as well as certain diseases.

As I wrote, though, I make some of my vegetables blanch, and I do it on purpose with no harm done. My blanching method for escarole plants involves nothing more than inverting clay flowerpots over their heads or crowding the plants close enough together so that neighboring plants force each other's outer leaves to be pushed up and over the inner ones; either way, low light levels make the leaves blanch. I've blanched celery and leek by mounding soil up against their stalks, and cauliflower heads by tying together their outer leaves or just snapping down one leaf to lay over the head. I've dug Belgian endive roots in fall and planted them in boxes brought down to my dark basement, where the roots pushed out pale, new sprouts. I've made cardboard collars to wrap around and keep light from cardoon stalks.

Even vegetables that are improved by blanching shouldn't be blanched willy-nilly. That chlorophyll is what harvests sunlight, converting it to food for plant growth. Young vegetables need to grow so can't afford to be starved for sunlight and, hence, food. Tender stems and leaves that result from blanching are more

Seakale leaves were blanched under clay pot.

prone to rot and insect attack, whether the plant is young or old. And fully grown plants need some energy just to stay alive.

All these caveats to blanching make autumn a good time of year to consider it. Leafy and stalky vegetables are fully grown by then so more growth is not needed. Cold weather slows down life processes so blanched vegetables can stay that way for weeks without expiring from lack of chlorophyll-plus-sun-producing energy. And that same cold weather also slows down insects and diseases, so that they pose less threat to succulent, pale stems and leaves.

Exceptions to autumn blanching would be a perennial vegetable. Seakale (*Crambe maritima*) is a cabbage relative that tastes best blanched. A clay flowerpot over the fat, perennial root just as it's sprouting new leaves in spring is all that it takes. I blanch and harvest seakale only for a while, early in the growing season, then uncover it so new leaves can fuel the roots for the next year's early sprout.

Blanching is not for all vegetables. Blanch a pepper plant and you'll end up with pale leaves and tasteless fruits on a severely stressed plant. Fruiting takes a lot of energy, for which a pepper plant needs plenty of sunlight.

Not even every leafy or stalky vegetable is improved by blanching. Blanch lettuce and it will be tasteless; blanch arugula and it will lack the zip for which we grow it. Cauliflower and celery are rarely blanched nowadays because self-blanching varieties—'Golden' celery and 'Snowball' cauliflower—have been developed. Even conventional celery is rarely blanched these days because most of us prefer the more robust flavor and texture of unblanched celery. White asparagus is now rare for the same reason. It's all a matter of taste (and texture).

ORGANIZATIONS

Wherein families migrate together around my garden, and for good reason

How many families am I having over to the vegetable garden this summer? I have to plan their seating arrangements.

I'm talking about plant families. An example of a plant family is the Mustard Family, known botanically as the Cruciferae, and including among its members cabbage, broccoli, collards, and Brussels sprouts. Their similarly pungent flavors and waxy, bluish leaves might also have earmarked them as being in the same family. Then again, the different parts eaten—the swollen stalks of kohlrabi, the leaves of cabbage, and the flower buds of broccoli—might indicate otherwise.

Most important in uniting this family, and the primary characteristic that unites members of any plant family, is the similarity of their flowers. All members of the Mustard Family have flowers with four equal petals in the shape of a cross. Hence, the name: crux is Latin for "cross," as in "Cruciferae" and "crucifixion."

Another prominent family that I'll undoubtedly have over this summer is the Leguminosae, better known as the Pea Family. This family also includes beans, and if I step out of my vegetable garden into my flower garden, lupines. On the way, walking across the lawn, I'll be stepping on another member, clover. Leaves of the Pea Family are usually made up of more than one leaflet, hence the 3- or, rarely, 4-leafed clover. But here, again, the characteristic

Flat-topped clusters of flowers of carrot, dill, and yarrow characterize the Carrot Family, Umbelliferae.

that most distinguishes all these plants as a family is their flowers. In this case, the flowers are irregular, having three different kinds of petals—two wing petals flanking an upper standard, and two lower keel petals.

The flowers of another family, the Carrot Family, are described by their botanical name, Umbelliferae. An umbel is a group of flowers, all of whose stalks radiate out from a common point atop a thicker stalk, resulting in a flat-topped or rounded cluster. Like an umbrella. Except for dill, which I grow for seeds and leaves, I rarely see the flowers of carrot, parsley, celery, parsnip, and other members of this family because I grow them only for their roots or leaves.

Five equal flower petals characterize one of the most-loved families in my garden, the Nightshade Family, botanically the Solanaceae. World famous members of this family include potato, tomato, eggplant, and pepper.

A plant family is characterized by more than just the number and shape of its flower petals. Taking a look at the flowers of cucumber, squash, melons, and pumpkins, I see that their flowers also have five equal petals. But the flowers of this family—the Gourd Family or, botanically, the Cucurbitaceae—are either male or female, and the central stalk of the female flower is capped by three stigmas to receive pollen. (Nightshade flowers all have both male and female parts, and female flowers have a single stigma.)

You may wonder: why all this fuss about plant families? Surely, the different families must be able mingle freely in the garden and get along. (After all, they're not human!) Yes, plant families can pretty much mingle freely.

The need for fussiness arises because members of a plant family usually share common pest problems. As examples, clubroot disease attacks the Mustard Family, blight attacks the Nightshades, and parsleyworms chew on leaves of the Carrot Family.

Except where it is sufficiently mobile or has an appetite for a wide range of families, a pest can usually be starved out by not planting members of a susceptible plant family in the same location more often than every three years. This is one of the rationales for "crop rotation." My vegetable garden is laid out in beds, with eight beds on each side of the main path running through the garden. One year a bed might be devoted only to tomatoes. Tomatoes are a no-no in that bed the next year, and the same goes for peppers or eggplants. That bed could be home to corn (Grass Family, Poaceae) or broccoli, cabbage, and kale. That year, the tomatoes get planted two beds away, as does the corn or Mustard Family the next year. And so on, year after year, different vegetables march like slow soldiers around the garden, two steps each year counterclockwise around the garden from bed to bed, with no family returning to where it previously grew for three years.

Crop rotation need not always be about pragmatism. Just for the fun of it, I've considered devoting (but haven't yet) a part of my garden to a single family, perhaps the Pea Family, for example, with a planting of lupines and sweet peas adding color to the dappled shade beneath a honeylocust.

Other Benefits of Crop Rotation

Crop rotation's measured benefits, benefits that go beyond pest control, make it hard to quantify the contribution of each. Perhaps the sum is greater than its parts.

Roots of different kinds of plants mine the soil for different spectra of nutrients, so crop rotation helps to maintain a balance of nutrients in the soil or, at least, to make more efficient use of fertilizers. The differing root morphologies of different kinds of plants likewise have different physical effects in the soil, each bringing its assets to the table. Contrast, for example, the relatively coarse roots of leeks with the fine and densely divided roots of peppers.

Biodiversity makes for a healthier soil and, in turn healthier plants. Introducing different kinds of plants on a regular basis (on a three or four year cycle with crop rotation) is one stab at maintaining biodiversity.

How plant families got put in order

It makes sense to base plant kinship on their sexuality, just as for humans and other animals. And the seat of plant sexuality is in the flowers. But it was not always so. A lack of understanding of plant sexuality prior to the 17th century was reflected in older systems of classification. For example, the third century BC Greek philosopher Theophrastus grouped plants as herbs, undershrubs, shrubs, and trees. Simple enough, but is a maple tree really related to a palm tree? Or a tomato to a marigold?

Credit for our present system of classification, based on flowers, goes to the Swedish naturalist Carl von Linne, who presented the system in a book, *Species Plantarum*, written in 1753. (Carl is usually known by the Latinized form of his name, Carl von Linnaeus.)

Commonalities among flowers of different members of a family are not always obvious. For instance, close observation is needed to winnow out similarities in the flowers of goldenrod and sunflower, both members of the Daisy Family, Asteraceae. This family is linked more by the intricacies of the small florets that make up their heads than by the sunny heads themselves. Two kinds of florets make up an Asteraceae flower head: ray florets, with petals that are asymmetric and fused into one long, strap-shaped "petal"; and disc florets, each having a relatively small, symmetric tube of fused petals. The head of a sunflower (most varieties) is mostly disc florets circumscribed by prominently petaled ray florets that create the decorative ring around the head. At the other extreme is a dandelion flower, all of whose florets are ray flowers, resulting in a powderpuff of yellow petals.

Linnaeus's system included what he called "classes," each determined by the number, proportion, and position of the stamens, the male flower parts. Classes were subdivided into "orders," each based on the number, proportion, and position of the pistils, the female flower parts.

Although Linnaeus's system was easy to use and gave all plants a convenient, binomial name, improvements were needed—and

Sunflower ray and disc florets.

Kinship within the Daisy Family, Asteraceae, including dandelion and sunflower, is evident when you examine individual florets.

made. In the 18th century, Bernard Jussieu found shortcomings in Linnaeus's system as he attempted to arrange in natural groupings plants at the Royal Gardens at Versailles. Why should male flower parts reflect a higher order of classification than female parts, anyway? So Bernard regrouped plants into more than a hundred orders that are now recognized as plant families.

All the older systems of classification were limited by their basis only on form and structure. As genetics and evolution became better understood in the 20th century, they were incorporated into the scheme of plant classification. Theories about plant relations continue to change with new knowledge about plant evolution and as new techniques, such as DNA fingerprinting, unravel the genetic makeup of plants.

Decisions will always have to be made as to which characteristics—whether petal number, a common ancestor, or similarities within a particular chromosome—are most important indicators of kinship. It's reassuring when a grouping jives with common experience: my daughter reaffirmed a common thread in the Carrot Family with her fondness for the tastes of celery, carrot, caraway, parsley, and dill. Even parsnip—yecch!

On Latin being a foreign tongue but providing a useful understanding of plant relationships

"Dawn Redwood" reads and speaks more easily than this tree's botanical name, *Metasequoia glyptostroboides*. But speaking the botanical name aloud slowly, meh-ta-see-KWOY-a GLYP-to-stro-boy-dees, is a delight to vocalize, a smooth dance of the lips, the tongue, and the upper palate.

Sensual pleasure aside, botanical names have a practical side. That woolly-leaved plant that sends up a candelabra of creamy yellow flowers each summer has a hundred or so common names. I call it mullein but other names include Aaron's rod, Adam's flannel, bullock's lungwort, and velvet plant. This plant has only one botanical name, though, *Verbascum thapsus*.

Each plant gets only one botanical name, and that name is recognized worldwide. A botanical name has two parts, both based on Latin. The first word in the binomial is the genus; each genus is subdivided into one or more species, the second word of the binomial. Genus and species names are always underlined or italicized. Genus is always capitalized and species is not capitalized unless it commemorates a person. Also, the species name is never written by itself; if the genus is obvious, it may precede the species in abbreviated form.

With the correct botanical name under my belt, if I want a packet of signet marigold seeds, I could ask for *Tagetes minuta* whether I'm in Andorra or Zanzibar. With plants like petunia, rhododendron, and fuchsia, at least part of the botanical name is the same as the common name in English. In some cases, using the botanical name might be the only way I could be assured of getting the plant I really want. If I wanted to plant bluebells, *Hyacinthoides*, and searched and asked for it by its common name, I could end with plants in the *Mertensia*, *Muscari*, *Campanula*, or *Eustoma* genus.

When I've planted lilies, I didn't want to pick from a grab bag of about 80 different species of *Lilium*; I wanted to plant—and did plant—the sweetly fragrant *Lilium candidum* (which actually is the only lily with the common name Madonna Lily). I also wasn't interested at that time in planting daylily (*Hemerocallis* spp.) or lily of the valley (*Convallaria majalis*), neither of which are true lilies. Neither is even in the Lily Family, let alone the *Lilium* genus!

A botanical name can tell a story about a plant: its discovery, its origin, its form. Adalbert Emil Redcliffe Le Tanneux von Saint Paul-Illaire discovered what became known as African violet growing amongst the lush vegetation in the mountain jungles of East Africa; in his honor the plant was given the botanical moniker *Saintpaulia ionantha*. *Juniperus virginiana* is a juniper native to eastern North America; *Juniperus chinensis* hails from China. Tulip tree, *Liriodendron tulipfera*, has a particularly descriptive name. *Liriodendron* was the ancient name for this plant, meaning "lily tree." And *tulipfera* means "tulip bearing." Tulip tree's flowers do, indeed, resemble tulips, as long as you don't look too closely. Tulip is a member of the Lily Family. Tulip tree is in the Magnolia Family.

Botanical names can shed light on plant relationships. The apparent similarity between peaches, plums, and apricots is confirmed in the similarity of their names: *Prunus persica*, *Prunus domestica*, and *Prunus armeniaca*, respectively. Knowing kinship can help decide what grafts might be successful. Grafts between the same genus and species almost always spell success. Success is even possible between different species of the same genus, which is why I plan to graft a couple of branches from my sweet cherry tree (*Prunus avium*) onto my tart cherry tree (*Prunus cerasus*). The sweet cherry doesn't bear reliably enough to warrant living here as a whole tree. Emulating George Washington, I'll chop the (sweet) cherry tree down following a successful graft.

In some cases, a species might be further subdivided into botanical varieties. For instance, cabbage and its close relatives are all *Brassica oleracea*. But cabbage itself is the botanical variety *capitata* (meaning head), Brussels sprouts are *gemmifera* (little gems), and broccoli is *botrytis* (cluster-like). The correct way to write the botanical name for cabbage, then, is *Brassica oleracea* var. *capitata*.

If you're put off by botanical names for plants, take heart because the situation used to be more awkward. Before the days of Linnaeus, baby's-breath was the botanical mouthful *Lychnis alpina linifolia multiflora perampla radice* and catnip was *Nepeta floribus interrupte spicatis pedunculatis*. Thanks to Linnaeus, the botanical names of these two plants now are the manageable and descriptive *Gypsophila elegans* and *Nepeta cataria*, respectively.

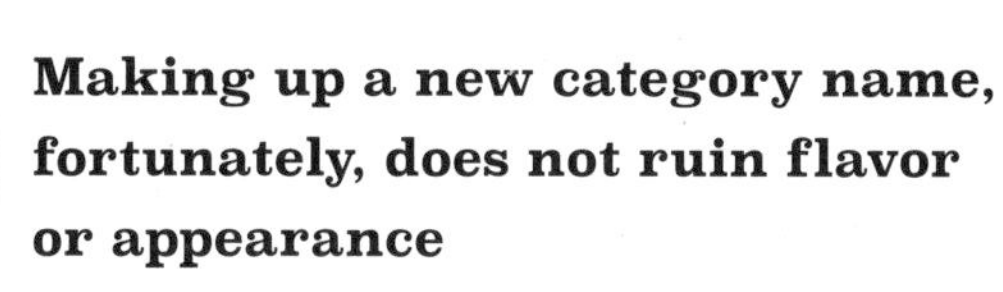

Making up a new category name, fortunately, does not ruin flavor or appearance

Mmmmm, how I like to bite into a cultivar. And look at the beautiful petals of a cultivar. And admire the autumn foliage of a cultivar.

A "cultivar?" What an ugly word for any plant with notable qualities.

A cultivar is any cultivated variety of plant. "Cultivated variety" contracts to "cultivar," a word that originated in the 1920s but didn't start to get traction until the 1970s. Some horticulturalists, myself included, avoid it. So even now, it's relatively unheard of. As I said, the word is just too ugly.

Before the word "cultivar" was invented, gardeners used the word "variety," but some people objected that this confused a "cultivated variety" with the previously described "botanical variety" (as in *capitata* being a botanical variety of *Brassica oleracea*).

Actually, species sometimes are pared apart to a level different from botanical variety. If a group of plants within a species is different from the rest, but not quite different enough to warrant "botanical variety" status, they may be placed within a "subspecies." And if they are a little more different than a botanical variety, they may become a botanical "form."

A cultivar (ughhh!) can be a botanical variety, subspecies, or form, or even an individual plant that is repeatedly cloned to make a whole population of identical new plants. For instance, 'Sunburst' honeylocust is, scientifically, *Gleditsia triacanthos* var. *inermis* 'Sunburst'. It's the yellow-leafed ('Sunburst') botanical variety of honeylocust (*Gleditsia triacanthos*) that lacks thorns (*inermis*). Botanical subspecies and varieties are always written in italics (or, in the age of of italic-print-lacking typewriters, underlined). Cultivar names are always enclosed within single quotation marks or preceded by the abbreviation "cv." without quotation marks; also, they are capitalized and, with exceptions, must be actual words in a modern language.

What distinguishes a cultivar is this: the group of plants has certain similarities, and the plants are intentionally cultivated. You might notice their similarities merely by sight, or the similarities might be more subtle, in the plant's physiology or chemistry.

The reason that you would intentionally cultivate a group of similar plants is because they have some desirable quality. Which is why I like to bite into a cultivar—of 'MacIntosh' apple, for example. A wild apple tree growing along the roadside is not a cultivar, unless I or someone else happens to like the fruit and starts making the tree into new plants. (The chances of a wild apple tasting as good as some cultivar is less than one in 10,000, though.) Similarly, 'October Glory' red maple has better autumn color than its run-of-the-mill, wild siblings. And I won't likely find a wild begonia with flowers as flamboyant as 'Rosebud Double Giant' begonia. If I do, I'll give it a name and I'll have a new cultivar.

The way that any cultivar is reproduced depends on the particular plant. Cloning, which I mentioned previously, is just one method. Many cultivars, such as 'Big Boy' and 'Big Girl' tomato, are F1 hybrids, the offspring from a deliberate mating of selected male and female parents. Some old varieties, errr...cultivars, of tomato, such as 'Belgian Giant' and 'Bonny Best', are self-pollinating, and reproduce true from seed now that they have been inbred for so many generations. (Plant, not human, generations.) Cabbages readily cross-pollinate, but an old cultivar like 'Early Jersey Wakefield' is maintained by growing it for seed in isolation.

All of which is not to say that all cultivated plants are cultivars. In some cases, a whole species is outstanding and worth growing, with no distinctive groups within the species. You don't find any cultivars of such garden-worthy plants as Father Hugo's rose, Virginia rose, or four o'clocks. Not yet, at least.

(I like many cultivars but hate the word, so I ignore the above rules except when needed to clarify whether a plant is a botanical or a cultivated variety.)

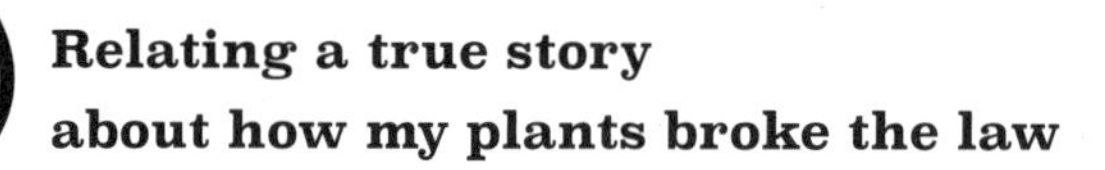

Relating a true story about how my plants broke the law

The intricacies of plant patenting were hit home for me with a shipment of strawberry plants one spring. The plants were of 'Chandler', a patented variety.

Strawberry plants send out runners, which are thin stems that sprawl along the ground and on the ends of which new plants form, which themselves take root and then also bear fruits and send out more runners and new plants. Those daughter plants that form at the ends of runners are useful for filling in a strawberry bed as well as for transplanting elsewhere to make a new bed.

Sirens. Flashing red lights. Whistles. Transplanting those rooted daughter plants would constitute a crime. But how about just letting the plants root by themselves? Propagation of any asexually (that is, not by seed) produced, patented plant is forbidden under the Plant Patent Act of 1930. The only exceptions are plants propagated by edible tubers—white potatoes, for example. (Not sweet potatoes, though, which are not tubers, but thickened roots.) Growers of white potatoes evidently were vocal enough back when the Act was being drafted to press for being able to save and replant their own potato tubers.

Some might argue that the Plant Patent Act was too long in coming. Had it been in place earlier, then Stark Brothers Nursery, which bought propagation rights to the original 'Red Delicious' apple for $3000 in 1894, would not have had to erect a cage around the original 'Red Delicious' tree. That cage only stopped people from using the original tree for propagation; once Stark Brothers started selling trees, though, those trees could be used by anyone to propagate new trees. The 1930 legislation was broadened, in 1970, with the Plant Variety Protection Act.

Seeds, which are sexually produced when pollen fertilizes eggs, can now also be protected by patents. So-called utility patents protect them, the same kinds of patents that can be used for everything from a new and better stapler to a dog whistle to—more recently and more controversially—genes. To be offered patent protection, a seed variety can't have been sold in the US for longer than a year, or elsewhere for longer than four years. The variety must also reproduce reliably and be distinct. Distinctiveness has always been a potential bone of contention, especially so since DNA fingerprinting can now be used to unlock a plant's genetic code, some of which is just "junk," not expressing any trait.

Patents are valid for 20 years (previously, 17 years), after which time anyone can propagate the plant for sale or otherwise. Someone could even then produce hybrid seeds, which are produced by crossing specific parents, because patents, available for anyone to see, spell out exactly how a product is made.

Enter trademarks. Whether or not a plant has been patented, it could be assigned a trademark name. What's more, that trademark is assigned to a company or individual, not to a specific plant. The company or individual could put that trademark name on any of their plants, even a few different ones. A patented variety also could be marketed under more than one trademark.

A patented plant is one thing and a trademark name is another. Patents have a limited life; trademarks can be renewed indefinitely, which makes trademarks very commercially useful. If you start selling some outstanding patented plant under a trademarked name, people will continue to purchase it under

Heritage birch breaks the law for plant names.

that trademark even after the patent expires, which is why plants are sometimes patented and then, right away, trademarked. Once the patent expires, the plant could be marketed under the original trademark or under other trademarked names assigned by other companies or individuals. Other people could propagate the patented plant, but not sell it under your trademark. Of course, if you or I recognize the plant only under its original trademark name, we'll seek out the plant under that name only.

A plant label stating "PPAF" (plant patent applied for) means, for plants, the same thing as "patent pending" for anything; "PVR" (Plant Variety Rights) means the plant has been patented. A plant may be patented, though, without it stating so on its label. Names of trademarked plants are followed by the symbol ® if the trademark has been granted a federal registration certificate. The symbol ™ is also valid and can be used by anyone claiming a trademark, without filing any paperwork.

Rules exist, of course, for patenting and trademarking, and I've recently learned that three birch trees I planted have broken the rules. They are Heritage birches. The variety name under which this plant was patented is Heritage; it was later trademarked Heritage. That's a no-no: a variety and trademark name must be different. Oh well, it wasn't me who broke the rule, and the plant is pest resistant and beautiful in spite of its brush with the law.

STRESS

On steps, human and otherwise, to avoid the havoc of icy cells during frigid temperatures

Not being able to don gloves and a scarf, or shiver, to keep warm, it's a wonder that trees and shrubs don't freeze to death from winter cold. They can't stomp their limbs or do jumping jacks to get their sap moving and warm up. The sap has no warmth anyway. Sometimes, of course, plants do succumb to winter cold. But usually that happens to garden and landscape plants pushed to their cold limits, not to native plants in their natural habitats or to well-adapted exotic plants.

Some plants—herbaceous perennials—opt for the easiest survival route, letting their tops die off each winter. Such plants anticipate frigid weather way back in late summer, when they start pumping nutrients and energy from their stems and leaves down to their roots. That's why asparagus stems go from lush green to tawny brown, and why coneflowers, delphiniums, and peonies are reduced, in autumn, to nothing more than a few dry stalks.

What's left of these plants spends a mild winter underground. Five feet down, the earth's temperature hovers around a relatively balmy 50 degrees Fahrenheit. Most plant roots don't run that deep, but even one foot down the temperature won't reach the minus 20 degrees Fahrenheit it could be outside. Practical Rule Number One: A thick blanket of some fluffy, organic material—leaves, straw, or wood chips, for example—limits penetration of cold into the ground and so keeps roots warmer through winter.

Low-growing plants, whose stems and leaves stay alive in winter, have it almost as good as those survived only by their roots. Near the ground, these plants aren't exposed to the full brunt of winter winds or cold. And if Nature decides to throw down a powdery, white insulating blanket, so much the better: those leaves and low stems are protected even more.

Practical Rule Number Two: In case Mother Nature is distracted with other activities, I provide my own low blanket for low-growing, woody or evergreen perennials—again, that fluffy cover of straw, leaves, or wood chips. Waiting to cover these plants until the weather turns relatively cold (soil frozen an inch deep is a good rule of thumb) lets plants acclimate to cold, and their stems and leaves have no chance of rotting beneath mulch that is still moist and too warm. Plants such as delphinium and cardoon are especially prone to rot, in which case an inverted flowerpot placed over the plant's crown, with mulch piled around and on the pot, provides an air space to prevent rot.

Think about it: water freezes at 32 degrees Fahrenheit—not a particularly cold temperature for a winter night—and plants contain an abundance of water. Water is unique among liquids in that it expands when it freezes, so you can just imagine the havoc that could be wreaked as water-filled plant cells freeze and burst. Yet plants that must stand tall all winter do, of course, deal with frigid weather.

One way trees and shrubs protect themselves from freezing is by shedding those parts most likely to freeze—their leaves. Which brings us to Practical Rule Number Three: I help plants along with their leaf shedding. No, not by pulling off the leaves, but by letting plants naturally slow down and begin to reallocate their energy resources, beginning in late summer. Reiterating the recommendation for getting plants ready for winter in the "Stems and Leaves" chapter, I don't stimulate them with excessive water, fertilizer, or pruning at that time.

Whether leafless or leafy in winter, trunks and branches of trees and shrubs do have to stand up and face the cold. Their living cells are filled with water. If this water freezes, the cells either dehydrate or suffer physical damage from ice crystals.

Water, whether in a plant cell or a glass, does not necessarily freeze as soon as the temperature drops below 32 degrees Fahrenheit. To freeze, water molecules need something to group around to form ice crystals, a so-called nucleating agent. Without a nucleating agent, water "supercools," remaining liquid down to about minus 40 degrees Fahrenheit, at which point ice forms whether or not a nucleating agent is present.

All sorts of things can serve as nucleating agents—bacteria, for instance—so plants may not be protected all the way down to minus 40 degrees Fahrenheit by having their water supercool. But winter temperatures don't plummet that low everywhere, so just a bit of supercooling may be all a plant needs to survive winter cold.

Plants have another trick for dealing with the cold, one that is effective well below that minimum supercooling temperature. That trick is to gradually move water out of their cells into the spaces between the cells. There, water can freeze without causing damage. Cell membranes are permeable to water, so as temperatures drop ice crystals that form outside plant cells grow with the water they draw out of the cells. With falling temperatures the plant becomes more threatened by dehydration than by freezing. Plants toughest to cold are those that are best at reabsorbing the water outside their cells when temperatures warm.

One other mechanism at work for the plants here is something called freezing point depression, which is why antifreeze keeps the water in your car radiator from freezing and why salt melts ice. Whenever you dissolve something in water, you lower the resulting solution's freezing point, more so the more that's dissolved. Plant cells are not pure water, and as the liquid in those cells losing water becomes more and more concentrated in sugars and minerals, its freezing point keeps falling.

Throughout this cold story, plants have not been passive players. In preparing for cold, cell walls strengthen and permeability of cells to water is actively altered. Here, again, I can step in: light supplies the energy that plants need to prepare for cold, so I make sure to site and prune plants so they get adequate light. Fruits are energy sinks, so I also make sure not to overcrop a plant, especially one that is borderline hardy. In addition to preparing plants for the cold, I can play around with microclimate, the climate right around a plant. (More on that later.)

Unfortunately, all this fiddling with a plant to help it through winter palls in the face of genetics. The very most that I can do to help trees and shrubs face winter is to plant those that naturally tolerate the coldest temperatures winter is apt to serve up.

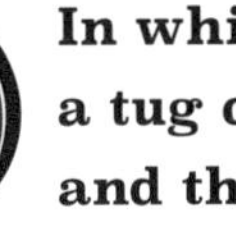

In which hot days bring on a tug of war between hunger and thirst, in plants

If it's not one thing, it's another. Pity plants in the heat of a hot, summer day! While I can jump into some cool water, sit in front of a fan, or at least duck into the shade, my plants are tethered in place no matter what the weather. And do not think that plants enjoy searing sunlight. High temperatures cause plants to desiccate and consume stored energy faster than it can be replenished. Stress begins at about 86 degrees Fahrenheit, with leaves beginning to cook at about 20 degrees above that.

One recourse plants have in hot weather is to cool themselves by transpiring water. Transpiration, which is the loss of water from leaves, can cool a plant by about 5 degrees Fahrenheit. Over 90 percent of the water taken up by plants runs right through them, up into the air, exiting through little holes in the leaves, called stomates. Carbon dioxide and oxygen, the gases plants need to carry on photosynthesis, also pass in and out through the stomates.

All this is fine provided there is enough water in the soil. If not, stomates close, transpiration and photosynthesis stop, and the plant warms. Even if the soil is moist, stomates might close in midsummer around midday if leaves begin to jettison water faster than the roots can drink it in. This situation puts most plants in a bind. Should they open their pores so that photosynthesis can carry on to give them energy, but risk drying out, or should they close up their pores to conserve water, but suffer lack of energy?

Enter cacti and other succulents (all cacti are succulents—that is, plants with especially fleshy leaves or stems—but not all succulents are cacti): their fleshy stems and leaves can store water for long periods. After more than a year without a drop of water, my aloe plant's leaves still look plump and happy.

Besides being able to store water in their stems and leaves, jade plants, aloes, cacti, purslane, and other succulents have another special trick, Crassulacean Acid Metabolism, for getting out of this conundrum. They work the night shift, opening their pores only in darkness, when little water is lost, and latching onto carbon dioxide at night by incorporating it into malic acid, which is

Purslane tolerates hot, dry weather
because of Crassulacean Acid Metabolism.

stored until the next day. Come daylight, the pores close up, conserving water, and malic acid comes apart to release carbon dioxide within the plant, to be used, with sunlight, to make energy. I have actually tasted the result of this trick in summer by nibbling a leaf of purslane—a common weed, sometimes cultivated—at night and then another one in the afternoon. Malic acid makes the night-harvested purslane more tart than the one harvested in daylight.

Another group of plants, called C4 plants, function efficiently at temperatures that have most other plants gasping for air and water. C4 plants capture carbon dioxide in malate, the ionic form of malic acid, which is a four-carbon molecule, rather than the three-carbon molecule by which most plants—which are "C3"—latch onto carbon. The enzyme that drives the C4 reaction is so efficient that C4 plants do not have to keep their stomates open as much as do C3 plants. The C4 pathway also does its best work at temperatures that would eventually kill a C3 plant, and cells involved in the various steps are partitioned within the leaf for greatest efficiency.

Pigweed, a C4 plant that is adapted to hot, dry weather.

C4 plants are indigenous to parched climates, but not uncommon visitors in gardens everywhere. Corn is a C4 plant. (Cool climate grains such as wheat, rye, and oats are C3 plants.) Looking at my lawn, I see another C4 plant. Hot, dry weather in August drives Kentucky bluegrass, a C3 grass, into dormancy. Not so for crabgrass, a C4 plant, which remains happily green. I also find some other C4 plants, in addition to corn, in my garden. As many vegetables and flowers flag, all of a sudden lambsquarters and pigweed, both C4 weeds (or vegetables, for those who like to eat them), appear as lush as spinach in spring.

Can I do anything to help out my plants in hot weather? Keeping the garden watered helps. Sprinkling or misting plants could keep them cool without their having to pull water up from the soil. But the 30 gallons of water that runs up through a tomato plant in a season, or the 50 gallons that flows through a corn plant, is for more than just cooling these plants. It also carries dissolved minerals from the soil into the plant. So it's debatable how well a plant would grow with too much misting. And besides, wet plants are predisposed to disease.

A better alternative to sprinkling plants is to grow plants adapted to the climate and the season. Lettuce, spinach, peas, and radishes are not the plants to grow for harvest in August. At least not in full sunlight; that's why, in midsummer, I grow lettuce in the shade beneath trellised cucumbers. Tomatoes, cucumbers, peppers, melons, and squashes, although they are neither cacti nor C4 plants, can take the heat. And, of course, so can C4 plants like corn, purslane, and vegetable amaranth.

Trellised cucumbers providing summer shade to lettuce.

No water, no matter—because I take these steps for drought

During the frequent dry spells of most summers, I tell my neighbor to thank me each time I water my garden. Why? Because my watering always seems to be followed by at least a passing thundershower. Perhaps similarly, a timely reading about how to cope with summer drought will ensure a season of abundant rainfall. Here goes (keep your raincoat and galoshes handy)…

I start preparation months ahead of potential dry conditions by turning my attention first to the soil. Any type of organic material—leaves, straw, compost, manure, sawdust, grass clippings—mixed into the soil will help plants weather dry periods. In sandy soils, organic matter acts like a sponge to hold water. In clay soils, organic matter opens up air spaces to promote far-reaching root systems. Caution: Churning soil does have its downsides (awakens weed seeds, destroys crystalline-like soil structure, disrupts earthworms and fungi, burns up organic materials) so should be done on a very limited basis, if at all. I gave it up decades ago.

During the growing season, these same organic materials laid on top of the soil as mulch prevent evaporation of water from the soil surface. If organic mulches are replenished each year, digging is never needed since earthworms and other creatures, rainfall, and gravity continually drag the lower portion of surface mulch down to mix into the soil.

What can I do with plants themselves to prepare for dry weather? If I were sure that every summer would be dry, I would grow only drought-resistant plants. This is not to suggest giving my garden over to the yuccas and saguaros of Arizona or to sugarcane, corn, lambsquarters, and other drought tolerant plants of moister, hot climes. Many familiar garden plants are at least somewhat drought tolerant. Such shrubs include juniper, potentilla, buckthorn (some buckthorns are considered invasive species), sweetfern, ninebark, and nannyberry viburnum. Amongst annual flowers, I could choose from cosmos, marigold, nicotiana, portulaca, sunflower, zinnia, and, interestingly, many of the annuals that are used for dried flowers, such as celosia, gomphrena, strawflower, and statice. Perennials that tolerate drought include yarrow, butterfly weed, coreopsis, poppies, coneflower, sedum, and baby's breath. Ornamental grasses such as pampas grass and blue fescue also are drought-tolerant (though most lawngrasses are greedy consumers of water).

No need to forsake vegetables to a dry summer. As mentioned previously, cucumbers, melons, okra, squash, and even tomatoes are drought tolerant, getting along with just enough water to plump up their fruits.

When water is limiting, decrease fertilization. More fertilizer means bigger plants and big plants need more water than small plants. Chemical (synthetic) fertilizers can exacerbate drought conditions by drawing water from plants in the same way that salty potato chips draw water from your lips.

If I couldn't water all plants sufficiently, due either to lack of time or water, I would first take care of those plants whose root systems have not yet reached far out into the soil. These plants include small transplants and newly planted trees and shrubs. Then I'd water other plants, beginning with those least able to tolerate drought.

What about the actual mechanics of watering? A rule of thumb is to water either infrequently and deeply (one inch or, equivalently, 0.6 gallons per square foot, once a week), or shallowly and frequently (same amount as the above but spread equally over the 7 days of the week). The latter is the essence of drip irrigation, whereby special water emitters drip water near plants at a specified rate. Since emitters (ideally) replace soil water at the rate that plants use it and only plants of choice get water, not paths and weeds, drip irrigation is very efficient. It's also easily automated, which is important since the water might need to be turned on and off one or more times daily.

No matter what type of watering system is used, tapwater can be supplemented with water caught from rooftops into barrels or cisterns. Some gardeners divert wastewater from their kitchen sink drains into the garden. When I plant a tree or shrub, I mound up the soil in small catchbasins around the base of the plant to catch water *in situ*.

A very local search for congenial weather

Cooler weather that slowly and steadily creeps into my yard each autumn prompts me to look around for microclimates. Microclimates are pockets of air and soil that are colder, warmer, more or less windy, even more or less humid than the general climate, due to the influence of slopes, walls, and pavement.

Every parcel of land, from a 40-acre farm field to a quarter acre lot, will have some microclimates, and siting plants with this in mind can spell the difference between whether or not they thrive or even survive. I'm banking, for instance, on the slightly warmer temperatures near the wall of my house to get my stewartia tree, which is borderline hardy here, through our winters. And I'm expecting spring to arrive early, with a colorful blaze of tulips, in the bed pressed up against the south side of my house. Proximity to paving also warms things up a bit.

But microclimate is not always about trying to keep a plant warmer in winter. It's also useful for keeping plants cooler. By training my hardy kiwifruit (*Actinidia* spp.) vines right up against the shaded, north sides of their hefty supports, I keep the sun off their trunks in winter and avoid the splitting that occurs when trunks are warmed during winter days, then precipitously cooled as the winter sun drops below the horizon. By planting the coveted blue poppy in a bed on the east side of my house, I hoped to give the plant the summer coolness that it demands. (That east bed was still too sultry; the plants collapsed, dead.)

Microclimates are important when growing fruiting plants that blossom early in the season because frozen blossoms do not go on to become fruits. Early season bloomers need microclimates that are slow to warm up.

South facing slopes stare full face at the sun, so these slopes warm up early in spring and are warmer in both summer and

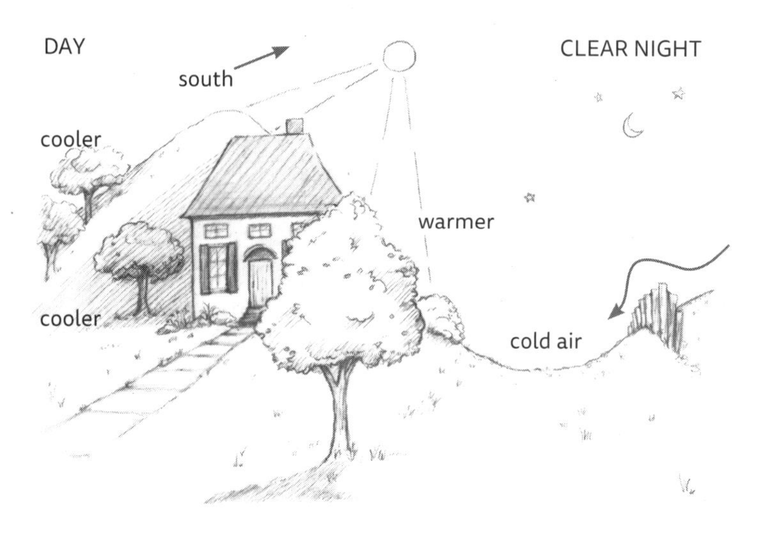

Effect of slopes and shadows on microclimates.

winter. Therefore, a south facing slope—even if the grade is only slight—can be used to hasten fruit ripening on a plant like persimmon, which blooms late but needs a long season when grown near its northern limits. Right after I push soil over the first seeds of sweet corn that I plant, I firm it over that hole with my foot at an angle to make a south-facing depression in the ground. That mini-slope will warm up just a wee bit sooner than flat ground.

The sun glances off north slopes, delaying their warming in spring and keeping them cooler in summer. Such a microclimate is ideal for an early blooming fruit tree like apricot or peach, and for plants, such as sweet peas, that enjoy cool summer weather. Likewise ideal for such plants is near the north side of a building, where shade remains through winter and the early part of the growing season.

If a slope actually has some elevation to it, the air is going to cool by one degree Fahrenheit every 300 feet going up the slope. If I had sloping ground, which I don't, and sought a cooler location for planting, I'd avoid planting at the very top of the slope, though, because the upper reaches are usually windy.

Counterintuitively, the very bottom of a slope will also be a cooler microclimate. On nights when the sky is clear, with no clouds or leafy trees to block re-radiation of the sun's heat from the ground back to the heavens, the air at ground level cools. An "inversion" occurs, with warmer air higher up. The cold air, which is denser than warm air, flows downhill to settle into depressions, just as a liquid flows downhill. A low point would be the worst possible location for planting strawberries, which grow near ground level and whose early blossoms are threatened by late frosts in spring. Riding a bicycle on a clear summer night, I've felt the effect of even a five foot dip in the road. Any dense fence or shrubbery on a slope stops the downward flow of cold air, which will pool, just as dammed water would, near the upper side of the barrier.

On chilly autumn mornings, I like to find microclimates for myself also: a sheltered, sun-backed nook near the brick wall of my house at which to sit with a cup of tea.

Seedlings' transition to the garden is helped along with tough love, timely and not in excess

Imagine that you had not set foot outside all spring...better yet, that you had spent all spring in a warmed cave...then tomorrow you went out and stayed there. At the very least, you would have to put your hands to your eyes for a while to shield them from the sun. And if the night was very cool—not unusual in spring—well, you would shiver. Fresh air and sunlight are great for the constitution, but you would have to first acclimate yourself to them.

The same goes, even more so, for vegetable and flower transplants. Indoors, where they get their start, they are, after all, coddled. They know nothing of wind, which can shake them up and dry out their leaves by too quickly drawing water from their stomata. Their tender cells know nothing about dealing with cool temperatures, or temperatures that swing 30 degrees Fahrenheit within 24 hours. Their leaves have yet to experience blazing sunlight.

Vegetable and flower transplants bought from a nursery would not necessarily fare any better. They've spent the better part of their youths in greenhouses, exposed to more sunlight than home-grown seedlings and, perhaps, the breeze from a fan, but still nothing compared to the great outdoors. The high humidity of a greenhouse does nothing to prepare them for drier air on the other side of the glass or plastic.

What's needed before setting transplants out in the ground is to have them undergo a process called "hardening off," which gets these plants acclimated to increasing intensity of sunlight, gusts of wind, fluctuating and cooler temperatures, and soil moisture levels that might border on drought one day and a week later turn boggy.

The kinds of changes that the lower temperatures of the hardening off period induce in coddled seedlings depend on the nature of the seedlings themselves. Seedlings of cabbage, lettuce, snapdragon, pansy, and other plants that can eventually stand up to temperatures below freezing develop a tolerance for cold by building up sugars in their cells. Cold also changes the composition of their cell membranes.

Tomato, marigold, zinnia, and seedlings of other plants that hardly tolerate temperatures below freezing begin to suffer from so-called chilling injury at temperatures below 50°F. Resulting changes in their membranes interfere with photosynthesis, causing a buildup, instead, of damaging toxins in their leaves. As a tomato or other warmth-loving plant becomes hardened off through gradual exposure to cooler temperatures, it becomes better able to repair and prevent such damage.

Once seedlings move outdoors, direct sunlight—whose intensity is as much as ten times that of light streaming through a sunny, south-facing window—can also do damage. Gradual exposure to more intense light, beginning in dappled shade or with just a few hours each day in full sun, thickens cell walls, fibers, and cuticles on both existing and new leaves. With increasing light exposure, chloroplasts, the green, light trapping energy factories in leaves, also move around and align themselves in such a way that the leaves turn darker green. And the stomatal pores of the leaves, through which water is lost and carbon dioxide and oxygen are exchanged, become more quickly able to open and close in response to changing conditions.

Hardening off needs to be gradual. Trying to toughen up plants too severely too quickly could send them into shock. Annual flowers and vegetables might respond by flowering prematurely. Flowering ruins a vegetable like celery, putting the brakes on stalk production and making those that remain too coarse too eat (fresh, at least; they're fine for soup). And you don't want flowers on a marigold plant before it has become big and bushy, or its growth will be stunted. The same goes for broccoli buds.

So what's needed is to find some cozy spot outdoors, a spot that's sheltered from wind and receives sun for only part of the day, or else dappled sun all day. I have a south-facing, brick wall along one side of my terrace that makes an L with a west-facing, white stucco wall; overhead is an arbor. The extra heat in that corner, the shelter from wind and all-day sunlight, and the part shade afforded overhead by the arbor make this corner ideal for beginning the hardening off of seedlings. About a week at that location starts the process.

Watering needs special attention because, barring rain or overcast conditions, plants are going to dry out much faster outdoors than they did indoors. Keeping plants slightly on the dry side makes for tougher growth, and gets them used to a condition they may have to experience once they are on their own in the ground. Too much water stress, though, could cause shock and its attendant effects.

I also pay attention to the weather. If a night threatens to be cold, I might, depending on the plants and the degree of cold, have to move the plants to a more protected spot outdoors—to a bench right up against the north wall of my house, for example—or indoors. (Fortunately, my garage/barn is on the other side of the brick wall of the L.) Despite the extra work, I move seedlings if there is any doubt about the weather. One cold snap could snuff out weeks of care, especially tragic if I'm growing 'Italian Sweet' pepper, 'Carmello' tomato, 'Lemon Gem' marigold, or other unique varieties that couldn't be replaced with ones from a local nursery.

After about a week, plants get moved to a more exposed location, one that takes just the edge off gusty winds and broiling sun. I continue to keep a close eye on watering and nighttime temperatures. A week at this second location and plants will be ready to be planted out at their permanent homes.

Young tomato plants protected against frosty night.

During the couple of weeks of hardening off, plant growth becomes slower and stockier. This is good; it shows that the plants are getting ready to face the world. Mother Nature can be fickle, though, so I stand ready to protect even these hardened off plants, once they are out in the garden, with overturned flowerpots or sheets if a late frost threatens. Ideally, this gradual hardening off, along with further protection, if needed, eases seedlings' transition to the garden so they hardly know they've been moved. Which is as it should be.

Unwanted plants—that is, weeds—are best understood before they are outwitted

"Well sir, all I can say is if I were a gate I'd be swinging! And if I were a watch I'd start popping my spri-ings! Or if I were a..." weed, well what would I do? With apologies to *Guys and Dolls* lyricist Frank Loesser, "Well Sir, if I were a weed I'd be seeding." Which is just exactly what pigweed does all summer—to the tune of over 50,000 seeds per plant! And dandelion, puffing into the air almost 20,000 seeds per plant! So one obvious way to lessen future weed problems is to rip out these plants before they spread their seeds.

But weeds are not so easily dispensed with. What else would I do if I were a weed? For one thing, I'd make sure that all those seeds I'm spreading would not sprout in unison. That way, there would still be plenty of seeds ready to sprout right after Farmer MacGregor comes through with his hoe to kill a bunch that already sprouted. And then some more for after he comes through again. Keeping a garden weed-free takes diligence.

If I were a weed, I'd also want my seeds to be able to rest in the soil for a long time, waiting for good growing conditions. And that's just what many seeds do that are buried within the soil. They lie there in a sort of suspended animation, just waiting for a bit of fresh air or light. I avoid awakening these weed seeds by never tilling or turning over my soil, which would expose them to the light and air they need to jar them awake.

If I were a weed, I'd also come up with ways of spreading my seeds as widely as possible. I might wrap my seeds in fruits, such as those of pokeweed or Virginia creeper, that are tasty to birds who would carry them long distances, inadvertently dropping seed "bombs" *en route*. Burdock seeds are carried far as their spiny burrs cling to animals' fur. Feathery appendages to milkweed and Canadian thistle seeds loft them into the wind to points unknown. Jewelweed, yellow woodsorrel, and creeping woodsorrel take a more proactive approach in spreading their seeds, each ripe fruit splitting open to forcibly eject its seeds a few feet from the mother plant. I keep out many such hopefully itinerant weed seeds by cleaning my wheelbarrow and tools of weedy debris before using them in the garden, and by maintaining a DMZ of mown lawn or mulched ground around my gardens.

But weeds can be sneaky. If I were a weed, I wouldn't rely only on seeds to get around. Creeping Charley rarely makes seeds but slithers in along my gardens' edges. It's easy to pull back and

Creeping woodsorrel, author's worst weed.

out, as long as it's done regularly. Creeping woodsorrel, likewise, creeps around (and spreads seeds!), its stems readily putting down roots wherever they touch soil. This weed is all the more difficult to control because its purplish leaves are camouflaged against dark-colored ground. Creeping woodsorrel is possibly the worst, but not necessarily the most abundant, of my weeds. (Because working the creeping stems out of the ground disturbs the plant to make it shoot seeds all over the place, I go at the clump with an open palm almost flat against the ground, then curl my fingers to lift up creeping stems and ejected seeds. Or I spray the plant with household vinegar to kill it *in situ* without disturbance; repeated spraying eventually starves the roots, and plant, to death.)

If I were a weed, I'd also want to be able to grow lustily, and to do so even under less than ideal conditions. When summer temperatures soar and weather turns dry, certain weeds come into their element. Most cultivated plants are gasping for water in the fullness of a hot, summer day and so must close their leaf pores, also shutting down photosynthesis and hence growth. Some weeds are succulents or have that C4 or Crassulacean Acid Metabolism, mentioned previously, that allows more efficient carbon dioxide uptake or uptake at night, when little water is lost through leaf pores. That's why purslane threatens to take over my vegetable garden and crabgrass my lawn in the heat of summer. Controls? I mow the lawn high so crabgrass seeds are shaded and have difficulty germinating. (Crabgrass is an annual plant.) Corn gluten is a natural herbicide that prevents seed germination; that could do the trick except that it also fertilizes the lawn with nitrogen, which means I'd have to mow more often. I control purslane and pigweed by eating them, or at least a portion of them. They are nutritious and tasty.

Which brings me to the finale: If I were a weed, I wouldn't be useful to humans for anything. Of course, if I were tasty or otherwise useful or sought after, I wouldn't be a weed. By definition.

A sometime threat that straddles the fence between living and nonliving

Winter cold ("cold" as in sickness) season is a good time to think about viruses. Plant viruses. Yes, viruses also infect plants. And these plant viruses have the same basic structure as the viruses that give us such diseases as the flu, mumps, and chicken pox. They're all nothing more than a single strand of DNA or RNA wrapped in a protein sheath and, in some cases, a lipid envelope.

Viruses occupy a twilight zone between the living and the nonliving. Outside of a host cell, viruses cannot grow, although some can "survive" for years in dried plant parts. Concentrate a solution of virus particles, and, just like table salt, they'll crystallize out. Only when a virus gets inside a cell does it come "alive" and multiply. But even then, it can do so only by commandeering its host's reproductive machinery.

Plants don't get fevers, runny noses, or achy joints, but have other ways of expressing their distress when infected with viruses. I've occasionally seen plants with leaves mottled pale yellow or crinkled. Or a plant might have become inexplicably stunted. Any of these symptoms could have been caused by virus infection. Symptoms depend on the type of virus and the type of plant. Virus symptoms might even come and go on a plant, resurfacing, for example, when cool temperatures slow growth. I once had an indoor lemon tree whose leaves were splotched with yellow every winter. A virus at work, perhaps. (Or a nutrient deficiency induced by poor root function in cold soil?)

A big difference in the way plants and animals react to viruses is that animals can fight off their infections, but plants cannot. No medicines (sprays) can help. Only in certain cases has human intervention helped, and that is when plants could, but viruses could not, tolerate a heat treatment lasting from a few hours to a few weeks. Sort of like curing a plant by giving it an artificial fever.

Lacking wings or legs with which to fly or crawl, or a mouth with which to chew itself into a plant, viruses need help with their evildoings. Even if they did have wings or legs, how fast could a "bug" move whose size is measured in nanometers—a unit a million times smaller than a millimeter? Viruses are adept hitchhikers, traveling from plant to plant in the gut of certain insects, nematodes, even fungi, then furtively slipping into a plant cell as any of these organisms begin feeding on a plant. This highlights one way that I prevent virus infections in my plants: I control insects or other organisms that spread viruses.

Removing weeds from near cultivated plants, and putting some distance between cultivated plants and related wild plants also lessens chances for virus spread. Aphids can carry cucumber mosaic virus from ragweed or chickweed to cucumbers, or raspberry mosaic virus from wild raspberries to cultivated raspberries.

And, of course, any time I know for sure one of my plants is infected with a virus, I immediately rip that plant out of the ground, hoping I have done so before the virus had a chance to spread disease to nearby healthy plants.

Viruses even solicit rides from humans. Tobacco mosaic virus infects a wide variety of plants, and smokers can inadvertently infect plants after handling tobacco products. Grafting a single, infected bud or stem onto a plant also spreads infection, as do pruning shears. I am very careful about my sources for scions for grafting.

Plant choice can help avoid virus problems. If I knew that bean mosaic virus was a problem on my green beans (it's not), I would plant a resistant variety such as 'Topcrop' or 'Provider'. Similarly, 'Latham' raspberry can carry raspberry mosaic virus with no ill effect.

Nurseries sometimes sell plants that are "certified virus-free." The nurseries produce virus-free plants from virus-infected plants by first taking a few cells from the tips of actively growing shoots, where virus infection has not yet reached. These cells are then multiplied, by tissue-culture, into a whole new plant, which can be the mother of virus-free progeny. After about 10 years, pretty much any raspberry planting picks up a virus from wild plants, although symptoms—such as a drop in yield or berries that are crumblier—are not always obvious. When my 'Fall Gold' raspberry's time is up, I don't dig up suckers from my old planting for my new planting, or from a neighbor's patch. I make sure to get certified virus-free plants from a nursery.

And even if a plant has a virus infection, that's not always a bad thing. In fact, it can be good, from a human perspective. Virus infections are what give the 'M.9' apple rootstocks on which I've grafted some of my apple varieties their ability to fully dwarf the tree; at maturity, these trees top out at only 8 to 10 feet high, with no apparent harm done by the virus. When this rootstock was cleared of virus infections—and given the new name 'M.9EMLA'—grafted trees grew 50 percent taller.

Tulips in paintings of the old Dutch masters show another sought-after virus infection. The pale petals of these tulips are streaked with darker colors, called color breaks, the result of virus infection. Virus infections are systemic, so would be passed along in bulbs as they are propagated.

Broken tulips contributed to the tulipmania rage in Holland in the 18th century. Then, the best bulbs sold for more than their weight in gold, with much of the trading involving not even the actual transfer of bulbs, but mere speculation. People gambled money, jewels, wine, even their homes. After four years the Dutch government intervened and banned further speculation. The virus does little harm to the bulb, affecting only flower color.

When a plant is having a problem, no reason to go ahead and blame it on viruses. Without a laboratory test, viruses are one of the most difficult plant problems to definitively diagnose. If a horrible sounding disease is needed on which to lay blame, I vote for verticillium or anthracnose instead? Or, perhaps the problem is just overwatering.

In which is clarified a name as a sign, rather than a symptom, of disease

Just as it does every year, summer brings a powdery white dusting to the leaves of my lilac bushes. Ah, if only this coloration were desirable, if we could affectionately refer to it as a silvery flecking or a sparkling sheen...but no, it's dull white and it's a sign of disease, a disease aptly called powdery mildew.

Your lilac's leaves also probably get dusted white with powdery mildew disease. And if you don't grow lilac, you may have had

Powdery mildew on peony leaves.

mildew on your cucumber's leaves, your rose's leaves, your gooseberry's leaves, or your phlox's leaves.

At first blush, all that powdery mildew could get me nervous about was its spreading from one plant to another. But I fear not, knowing that many different fungi are responsible for the disease we collectively call powdery mildew, and each of these fungi have specific plant hosts. Thus, the fungus that causes powdery mildew on lilac (the fungus is *Microsphaera alni*) can't cause powdery mildew on rose (caused by the fungus *Sphaerotheca pannosa*). Even though powdery mildew on cucumbers and phlox are caused by the same genus and species of fungus, *Golovinomyces cichoracearum*, the race of this fungus that attacks cucumber is incapable of attacking phlox; and *vice versa*.

Powdery mildew, like some other common names of plant diseases, describes a sign of the disease, the "sign" being the look of the actual disease causing organism. In the case of powdery mildew, that sign of disease begins with white patches on leaves that, under suitable conditions, eventually coat whole leaves. The whiteness is the actual disease-causing fungus. The fungal threads spread primarily on the outside of the plant and periodically send "pegs" down into the plant with which to extract food. As the disease progresses, leaves become deformed, infected fruits turn leathery, and flower buds fail to open.

Other descriptive, self-explanatory common names for diseases are scab, leaf spot, and brown rot. These names describes "symptoms," that is, what a plant's response to the disease looks like rather than the look of the disease-causing organism itself. Anthracnose, a name that doesn't readily conjure up an image of a symptom, is a name for diseases that result in distinct dead areas in leaves.

Aptly named "apple scab" on apple fruits.

Diseases having the same descriptive name are not necessarily all caused by even the same kind of "germ." Blight of pear, for example, is caused by a bacterium, while blight of peony is caused by a fungus. Both diseases have "blight" in their names because the term blight is generally used for any disease that causes a plant or plant part to suddenly die.

Descriptive names for plant diseases can get more specific. The blight attacking pears is called fire blight, so-named because infected stems look as if they have been singed by fire. The peony affliction is called botrytis blight, "botrytis" meaning grape-like and referring to the grape-like clusters of fungal spores seen under magnification.

Figuring out what a disease is called makes for better gardening because it forces a closer look at just what's going on. When I see something amiss with a plant, I first note what part or parts are affected. Then I try to give the symptom a name—Joe, Ed, Ted, no no, just kidding; but rather rust, rot, dieback, canker (a localized area of dead bark, often dark and sunken), anthracnose, leaf spot, etc. Closer inspection might bring on even more descriptive names—brown rot or black rot, as examples or, even better, scientific names. (A good book on plant diseases helps at this point.) Look at the ugly spots on my tomato plants' leaves. *Alternaria solani* (early blight) marks leaves with brown, round spots each a half-inch in diameter and surrounded by concentric

Early blight disease on tomato leaf.

Septoria leaf spot disease on tomato leaf.

rings. *Septoria lycopsersici* (septoria leaf spot) causes spots that are small, round, and gray, each surrounded by a single, dark margin. And *Phytophthera infestans* (late blight) causes greenish-black splotches. In the case of tomato leaf spot diseases, the control for all is the same: a site where sun and breezes quickly dry leaves, crop rotation, and thorough garden cleanup in the fall.

In contrast to most other fungal diseases, powdery mildews thrive in dry weather. Rainfall actually washes the spores off plants. The spores do need a little moisture to induce them to germinate and infect a plant, and dew left on leaves when cool nights follow hot days is sufficient.

As with other fungal diseases, the quicker plant parts dry (in this case from dew) the less time available for fungal spores to germinate. A good site and pruning facilitate drying. Pruning perennial plants before growth begins the following season also helps by getting rid of infected plant parts in which the fungus can spend the winter and lie in waiting, given rain and warmth, to infect the following year's plants.

Good nutrition and natural resistance are part of the arsenal against fungal diseases. Plants are more resistant to powdery mildew if they are adequately nourished with potassium and not overly lush from too much nitrogen. 'Miss Lingard' ('Wedding Phlox') and 'Eva Cullum' are some few phlox varieties resistant to powdery mildew; 'Miss Kim' is a resistant lilac; and 'Poorman', 'Hinonmakis Yellow', and 'Red Jacket' are resistant gooseberries.

If site selection, pruning, and fertilization fail to thwart powdery mildew, sprays can come to the rescue. Before spraying, I weigh the benefits of spraying against the costs in terms of time and trouble. Powdery mildew on my lilacs and zinnias usually occur late enough in the season to do the plants little harm, so that the effect is purely cosmetic, which I can ignore. If I do decide to spray, I choose a fungicide with care, because some commonly used fungicides—Captan, for example—are ineffective against powdery mildew.

Some organic sprays are quite effective against powdery mildew. One is baking soda, prepared by mixing four teaspoons of baking soda in a gallon of water, adding two tablespoons of soybean oil to make it stick better to leaves. Baking soda is sodium bicarbonate; commercial formulations, containing instead potassium bicarbonate (Kaligreen® and Greencure®, for example), are less likely to damage plants.

Two other organic sprays for controlling powdery mildew are powdered sulfur and horticultural oil. Certain plants are damaged by sulfur, and most plants are damaged by either sulfur or oil if sprayed in very hot weather, when temperatures are above 90 degrees Fahrenheit. Neither material should be sprayed within 2 weeks of the other. "Damage this" and "damage that" notwithstanding, both sulfur and oil are relatively benign to plants and to non-target creatures, such as you and me.

Powdery mildew is no newcomer to gardens. It is likely the mildew mentioned in the biblical passage "I smote you with blight and mildew; I laid waste your gardens and vineyards..." (Amos 4:9) because of its prevalence, in contrast to most other fungal diseases, in the hot, dry climates of the Mideast. Similarly, sulfur has a long tradition of use as a fungicide, being mentioned as far back as 800 BC in the epic poems of Homer. Recently, both powdery mildew and sulfur have become unfortunate traditions in my garden, on grapes and gooseberries.

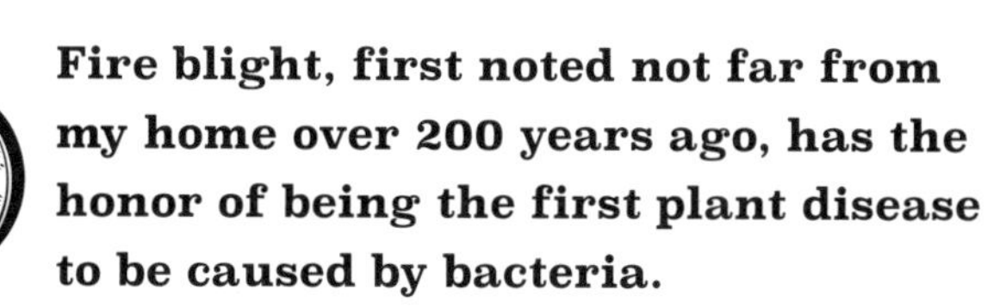

Fire blight, first noted not far from my home over 200 years ago, has the honor of being the first plant disease to be caused by bacteria.

Would-be pear growers often are cautioned against planting pears, with the admonition that fire blight lurks in the air, waiting to spell death to the trees. This disease, caused by a bacterium, was first reported in 1793 in the Hudson Valley of New York (not far from where I grow pears) by William Denning, in the "Transactions of the New York Society for the Promotion of Agricultural Arts and Manufacturers." Soon after it was first reported, treatments for fire blight and theories about its cause abounded. The

Pennsylvania Horticultural Society offered a 500 dollar prize in 1837 for "the person who shall discover and make public an effective means of preventing the attack of...fire blight." Rusty nails in the trunk, soapsuds on the ground, and brimstone rag wrappings were among the suggested cures. Early theories attributed fire blight to insects, frozen sap, sun-scald, and fungi.

At the end of the 19th century, Thomas Burrill at the University of Illinois determined that a bacterium was the cause of fire blight. This discovery gave fire blight the distinction of being the first plant disease recognized to be caused by a bacterium. The bacterium was officially named *Erwinia amylovora* in the 1920s.

There's no denying that fire blight can be a serious disease of pears—often killing whole trees—but it is a disease that can be controlled with a little forethought and a watchful eye. No need even to spray.

Fire blight is a disease aptly named, for afflicted plants look as if they have been singed with a blowtorch. Leaves blackened but stuck to branches, and tips of tender young shoots curled in a "shepherd's crook" are sure symptoms of the disease. Afflicted tissues often exude little droplets of bacterial ooze. (Superficial black smudges on leaves and twigs are due to "sooty mold," sometimes mistaken for fire blight, but actually fungi growing on honeydew dripped onto the surface of leaves by insects like aphids and pear psylla.)

Characteristic symptoms of fire blight disease.

The bacteria become active in the spring, as they are awakened by moisture and warmth from their overwintering cankers. Wind, insects, and rain then spread bacteria to succulent new shoots, blossoms, fruits, and leaves. Hail damage makes trees especially susceptible to infection. The infection cycle continues as long as temperatures are warm and humidity is high, which is one reason that, although fire blight is widespread, it is most (justifiably) feared in the Southeast. When summer has ended and tree growth slows, the bacteria once again become dormant. By this time, the bacteria have grown towards the trunk killing twigs, branches, perhaps the entire tree.

Now for the counterattack. Pruning shears are a useful weapon. In fact, a watchful eye and diligent pruning can completely check this disease.

Once in winter and periodically throughout the summer, cut off branches and twigs a foot below any infection. Diseased tissues should be pared away from limbs too large to remove. Sterilize pruning shears with either a dip in, or a wipe, with alcohol, Listerine®, Lysol®, or Pine-Sol® between cuts during summer pruning to avoid spreading the disease. While you're at it in summer, snap off vigorous suckers thrown up from branches and the rootstock; such succulent growth is particularly susceptible to fire blight infection. No need to sterilize tools between cuts in winter.

Because fire blight bacteria most readily attack succulent stems, restraining tree growth is another means of averting the disease. Excess nitrogen fertilizer or over-zealous winter pruning stimulates lush growth in plants. A general guideline for fertilizing pears is to use about an ounce of actual nitrogen per year of age of the tree, to a maximum of 1.25 pounds per tree. The amount of actual fertilizer needed depends on its percentage of nitrogen. For example, if the nitrogen source is soybean meal, which is 7 per cent nitrogen, about three-quarters of a pound would be needed per year of age of the tree, up to 15 pounds. I fertilize my pear trees with compost, which provides many benefits in addition to its one per cent nitrogen; 5 pounds of compost per year age of tree would be the equivalent recommendation. New stems on a young, healthy pear tree should grow no more than a foot or two per year; on an older tree, no more than a half-foot to one foot.

Lawn under my pear trees acts as a safety valve to help regulate tree growth. If pear trees are growing too fast, I let the lawn grow long to soak up excess water and fertility. Not enough growth on a pear tree? Then I mow the lawn closer and more often. But grass should not be allowed to grow right up to the trunk of very young trees, or their growth will be stunted.

Not all pear varieties are equally susceptible to fire blight. The popular Bartlett, Bosc, and Anjou pears all are very susceptible. The small, sweet Seckel pear is resistant.

In the early part of this century, the U. S. Department of Agriculture embarked upon a breeding program to develop pears resistant to fire blight. Such early introductions as 'Orient' and 'Pineapple' are fire blight resistant hybrids of Asian and European-type pears and are best suited for cooking. More recent introductions, like 'Magness', 'Maxine', and 'Moonglow', are fire blight resistant and high quality for fresh-eating.

I grow over 20 varieties of pears, and—knock on wood—have never seen fire blight here. I grow 'Magness', 'Maxine', 'Harrow Delight', and 'Seckel', which all are naturally resistant to fire blight. Many of my other trees, including 'Clapp's Favorite' and 'Aurora', are not resistant to fire blight, so I also monitor all the trees' vigor and keep an eye out, pruning shears in hand, for any symptoms of disease. Last, but not least, I am careful not to introduce any scions or plants that carry infection. Fire blight affects over 130 members of the Rosaceae (Rose Family), in varying degrees. Apple, quince, mountain ash, cotoneaster, and hawthorn are some of the prominent family members that can be attacked.

As a backyard gardener, I have the option of growing any one of the over five thousand existing pear varieties (only two dozen planted so far, though). The best varieties of pears—picked just as the fruit begins to slightly soften and yellow, then ripened indoors—are buttery, juicy, and intensely aromatic. Threat of fire blight has not denied me that pleasure.

Since fire blight was found to be caused by bacteria, other bacterial diseases have been identified. Among some other common ones are cucumber wilt, crown gall, and bacterial canker and bacterial leaf spot of peaches. As with fire blight, they can be kept at bay by avoiding the introduction of infected plants, seeds, or soil; by the use of commercially available antagonistic organisms or their products; by spraying copper compounds; by controlling insects that carry bacteria from one plant to the next; and, best of all, by choosing varieties of plants that are naturally resistant to specific problem bacteria.

Generally, though, I welcome and nurture bacteria in my garden. Among the, literally, millions of species of bacteria, most are either innocuous or else beneficial. Thankfully, because they are around and within us in large numbers. Gardenwise, just a single teaspoon of garden soil may be home to 1 billion of them.

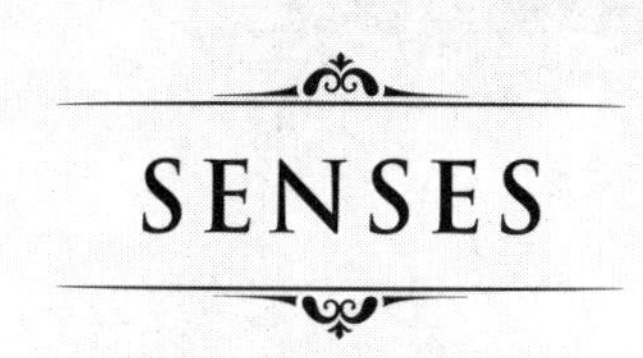

SENSES

In which I elucidate, abet, and alter the color of leaves, vegetables, and flowers

Fall Color

The sugar maples are glorious each fall, but not all equally so. I'm lucky to have growing behind my yard one of the best I've seen. Not only is its color especially fiery but the tree remains full with leaves after other maples have dropped most of theirs. I'd like to see more maples looking like this one—a not impossible dream.

Knowing what puts color in leaves opens up the possibility for ratcheting it up. Green is from chlorophyll, most welcome in spring and through summer, but not what interests me in fall. Chlorophyll must be continually synthesized for a leaf to stay green. The shorter days and lowering sun of waning summer are what trigger leaves to stop producing it, unmasking other pigments lurking there.

Leaves' yellow and orange colors are always present, thanks to carotenoid pigments, which help chlorophyll do its job of harvesting sunlight to convert into plant energy. I offer thanks to carotenoids for the warm, yellow glow they give to gingko, aspen, hickory, and birch leaves.

Tannins are another pigment, actually metabolic wastes, that all summer are hidden by chlorophyll. Their contribution to the fall palette are the season's subdued browns, notable in some oaks and enriching the yellow of beeches.

Because leaves harbor carotenoids and tannins all summer long, nothing particular about autumn weather should either intensify or subdue their autumn show. The only glitch could be an early, hard freeze that occurs while leaves are still chock full of chlorophyll. In that case, cell workings come abruptly to a halt and all we're left with is frozen, green leaves that eventually drop without any fanfare.

Autumn color also spills out reds and purples, most evident in red maples and some sugar maples (like mine), scarlet oak, sourwood, blueberry, and winged euonymus. Those reds and purples come from yet another pigment, anthocyanins. Except for trees like 'Purple Fountain' beech and 'Royal Purple' smokebush, whose leaves unfold dusky red right from the get go in spring and remain so all season long, in most leaves anthocyanins do not begin to develop until autumn.

Anthocyanin formation requires sugars so anything that I or the weather does to promote sugar accumulation in autumn will increase anthocyanin levels in leaves. The weather's role is to offer warm, sunny days to maximize photosynthesis, and cool, but not frigid, nights to minimize nighttime burning up of accumulated sugars. A cloudy, rainy autumn means less red because less anthocyanin is formed, and any that does form is diluted.

My role in ratcheting up the reds and purples is to make sure that leaves bask in light. I plant a tree where light is adequate (for that species) and, as necessary, prune so that branches don't shade each other. Street lights don't count as light, and actually have a negative effect, disrupting the signal that days are getting shorter and it's time to slow chlorophyll production.

My other role in the autumn show is to plant trees genetically programmed for good autumn color. And among the most colorful-leaved trees and shrubs—which, besides those previously mentioned, include goldenrain tree, hickory, ironwood, black tupelo, and fothergilla—individuals within each species might pack a bigger wow than the others. Hence the spicebush variety 'Rubra', brick red in fall, or 'Wright Brothers' sugar maple, whose leaves become a mottling of gold, pink, orange, and scarlet.

All trees of a named variety are genetic replicas of each other, so must be reproduced by cloning. Cuttings are one method of cloning, but many trees do not root readily from cuttings. To have a genetic replica of my especially fiery sugar maple, I could create a new one by grafting a stem from it onto any sugar maple seedling that happened to pop up in the woods or my garden. I might just do it.

Vegetable and Flower Color

At four years old, my daughter was very fond of green beans; I decided to shake things up a bit and grow 'Royal Burgundy', a purple-podded "green" bean. As she and I plucked pods from the plants, she turned to me and asked why cooking turns the pods green. Which it does.

Of course, I couldn't, at that time, say, "Anthocyanins, my dear." (That time came six years later.)

As with autumn leaves, anthocyanins are what put the purple in purple green beans, as well as in purple broccolis, grapes, and plums. Anthocyanins are also what make roses and geraniums red, cornflowers and delphiniums blue, and bigleaf hydrangeas either blue or pink. And again, as with autumn leaves, anthocyanins in these plants have no hand in their yellows and oranges; those colors come from carotenoids, which also are responsible for certain reds. In the case of beets and bougainvilleas, the red comes from yet another pigment, called betacyanin.

How can anthocyanin make one vegetable or flower red and make another vegetable or flower blue? Or, in the case of bigleaf hydrangea, either pink or blue.

Within plant cells, anthocyanin molecules stack together, and how tightly they stack determines their color. Tight stacking makes blue; loose stacking makes red or pink. Acidity of the cell sap, which is genetically programmed in a plant, affects anthocyanin stacking. Anthocyanin that is red in the very acidic sap of a rose petal is blue in the less acidic sap of a cornflower petal.

Anthocyanins change color with changes in the acidity of their environment, and eventually turn colorless as their surroundings turn more alkaline.

Two things happen during cooking to make 'Royal Burgundy' beans turn from purple to green. A direct effect of the heat is to cause decomposition of anthocyanin. Less anthocyanin, less purple. An indirect effect of heat is to burst apart cells, diluting the acidity of the cell sap. The green color, which was present but masked by the anthocyanin, becomes prominent once the anthocyanin concentration drops and what survived the cooking is bathed in liquid insufficiently acidic to make it purple.

A similar thing happens when cooking red cabbage. It turns colorless after a while, and the same can be expected from purple broccoli, purple asparagus, purple tomatillos, even purple peppers. With carotenoids giving them their flair, red peppers stay red through cooking.

Except to entertain my daughter, it perhaps seems foolish to grow a purple vegetable that's anyway going to turn green after I cook it. It's not as if that purple color does much for flavor. In nature, though, anthocyanins do have a purpose, and that is to attract insects to flowers, to allow for photosynthesis at colder temperatures and lower light conditions (hence the purple or red colorations of many alpine and arctic plants), and to protect plants from ultraviolet radiation. Protection from sunlight is why the shoulders of my carrots, exposed to sunlight, turn purple.

Actually, the purple in a vegetable does do me some good. The color spruces up the look of the vegetable garden. Also, because foliage of 'Royal Burgundy' stays green, I can more easily pick out the purple pods from among the leaves. And perhaps birds can more easily spot and devour green cabbage worms on purple broccoli and red cabbage than on green varieties of either vegetable. (Not in my garden.)

Furthermore, those purple beans don't really have to turn green before they're ready to eat. My daughter enjoys green—or purple—beans raw. And I can prevent, or lessen, the color change of any cooked, purple vegetable by soaking it prior to cooking in vinegar or lemon juice to increase the acidity. Then minimize cooking. It's all for show because of the little, if any, effect of anthocyanins on flavor. No matter in the case of 'Royal Burgundy', though, because we never thought they tasted very good anyway.

Returning to bigleaf hydrangeas whose flowers (botanically, they aren't flowers, but bracts, which are modified leaves) can be either pink or blue. My choice. Your choice. The way to effect this change is by changing the acidity of the soil.

With hydrangea, the effect of soil acidity on color is indirect, via aluminum, which becomes more soluble and increasingly absorbed by plants in increasingly acidic soil. Once transported to the hydrangea bracts, aluminum neutralizes charges on anthocyanin molecules so they can stack more closely, resulting in blue coloration. Soil acidity and the resulting aluminum uptake could also affect the pink to blue color change with other flowers. Except that aluminum is toxic to plants, and toxicity usually results at levels needed to effect that color change. Hydrangeas are not immune to aluminum toxicity. For blue hydrangeas, the ideal pH is in the range 5.0–5.5.

So all that's needed to grow bigleaf hydrangeas with blue flowers in soils that are naturally alkaline or just slightly acidic is to acidify the ground with sulfur or aluminum sulfate. Sulfur, a naturally occurring mineral, oxidizes in the soil to become sulfuric acid, and the resulting acidity increases plant uptake of aluminum that is naturally present in most soils. Adding aluminum sulfate both acidifies a soil and adds aluminum.

For pink hydrangeas in a soil that is naturally very acidic, limestone is the ticket. The ideal pH range, in this case, is 6.5 to 7.0. For quickest effect, the limestone needs to be dug into the soil at planting time.

While the genetically programmed color expression of anthocyanins is affected by acidity within plant cells, cell acidity is unrelated to soil acidity. What's more, the effect of acidity within the cells is opposite from the effect of acidity in the soil: increased cell acidity spreads anthocyanin molecules further apart, so the color is pink; increased soil acidity puts enough aluminum in the cells to collapse the anthocyanin molecules, so the color is blue. So while cell acidity determines the color of many flowers, you can't affect that acidity.

Some flowers, such as wild morning glories, are pink when they initially unfold, then turn blue. Why? No one knows—yet.

An Italian who tied together plant growth, art, and other things too innumerable to mention

Although not obvious, a pine cone, an Italian mathematician, and Manet have something in common. I could even go a step further and state that this mathematician—Leonardo of Pisa, living in the 13th century—shares something with almost the whole plant kingdom which, in turn, also shares something with much of art itself, not just painting. The mathematician, better known by his *nom de plume*, Fibonacci, knits these diverse arenas together with a number sequence he discovered, a sequence that keeps popping up in art and nature.

Spiraling bracts of a Norway spruce cone.

A Fibonacci sequence is simple enough to generate: starting with the number one, merely add the previous two numbers in the sequence to generate the next one. So the sequence, early on, is 1, 1, 2, 3, 5, 8, 13, 21, and so on. Beginning in the 17th century, some mathematicians and, most famously, in 1843, Jacques Binet, spared us the tediousness of generating a really large Fibonacci number by coming up with a formula for it instead.

I step outside to find an intact pine cone (or any other cone). Looking carefully at the cone, I notice that the bracts that make it up are arranged in a spiral. Actually two spirals, running in opposite directions with one rising steeply and the other gradually from the cone's base to its tip. Counting the number of spirals in each direction is a job made easier by dabbing the bracts along one line of each spiral with a colored marker. The number of spirals in either direction is a Fibonacci number. I just counted 3 and 5 on a Norway spruce cone.

If cones aren't readily available or the weather outside is too miserable, I can find Fibonacci somewhere indoors—like on a pineapple. By focusing my eyes on one of the hexagonal scales near the pineapple's midriff, I can pick out three spirals, each aligned to a different pair of opposing sides of the hexagon. One set rises gradually, another moderately, and the third steeply. Counting the number of spirals, I find 8 gradual, 13 moderate, and 21 steeply rising ones. Fibonacci numbers, again.

Scales and bracts are modified leaves, and the spiral arrangements in pine cones and pineapples reflect the spiral growth habit of stems. This can be confirmed by bringing a stem from some tree or shrub indoors for a look at its leaves or, in winter, at its buds, where leaves were attached. The leaves or buds range up the stem in a spiral pattern, which keeps each leaf out of the shadow of its neighbor just above it. The amount of spiraling varies from plant to plant, with new leaves developing in some fraction—such as 2/5, 3/5, 3/8, or 8/13—of a spiral. Eureka, the numbers in those fractions are Fibonacci numbers!

I determined the fraction for the dormant stem I was holding by finding a bud directly above another one, then counting the number of full circles the stem went through to get there while generating buds in between. So if the stems made 3 full circles to get a bud back where it started and generated 8 buds getting there, the fraction is 3/8, with each bud 3/8 of a turn off its neighbor upstairs or downstairs. I just computed fractions of 1/3 and 3/8 on a single apple stem; the fraction can change as any plant's stem grows, but remains a fraction involving Fibonacci numbers.

I haven't forgotten about Manet and the artists. It turns out that there are certain proportions that we humans universally find pleasing—the rectangular proportions of a painting, for example, or the placement of a focal point in a painting. Enter the Golden Ratio, which, for the latter example, states that the ratio of the distance from the focal point to the closer side of the painting, to the distance from the focal point to the farther side of the painting, is the same as the ratio of the distance from the focal point to the farther side of the painting, to the painting's whole width. A pleasing ratio, it turns out, is always 0.618… or, if you want to use the inverse, 1.618…. (No other number and its inverse can match digits like that, incidentally.) Enter Fibonacci: divide any Fibonacci number by the Fibonacci number before or after it and you get 0.618… or 1.618…, not exactly at first, but closer and closer the higher the Fibonacci number you start with. Try it.

Of what use could Fibonacci be in the garden? It might be used in design—gauging a pleasing ratio of potted plant height to the height of its decorative container, or the layout of a knot garden. (I admit to just "eyeballing" such things and what turns out to look nice is what Fibonacci would have ordered.) I might "use" it to appreciate the orderliness of plant growth, as manifest in the spiraling arrangement of buds or leaves along a stem, or the seeds within a sunflower head.

Here I make sense of scents, equally so for insects and humans

Wave after wave of scent wafts across my terrace as the garden awakens in spring. Most prominent in my "back forty" are the aromas from daffodil blossoms, followed by those of plum, flowering currant, and then dame's rocket, pinks, and roses. Olfactory pleasures, like the other sensual pleasures that flowers afford us, are incidental to the flowers. Evolutionarily speaking, we don't return the favor with anything more than the carbon dioxide that we—and every other animal—breathe out.

Rather than smelling pretty for us, flowers do so to attract pollinators. As such, floral aromas might mimic countless other kinds of aromas, depending on just what creature a particular flower is trying to attract.

Because floral aromas are not there to please us, not all of them smell good to us. Skunk cabbage is a good example, but there are much worse—or better—examples. The arum lily of South Africa, for example. From its spathe, a spikelike inflorescence of many small flowers rising up from what looks like an upended purple skirt, wafts the smell of rotting flesh. This aroma is just the ticket for attracting the carrion beetles that pollinate this plant. Heat generated at the interior of the inflorescence heightens the morbid effect, besides helping pump the aroma out into the atmosphere.

On to more pleasant aromas... to flowers that mimic pheromones, which are scents that female animals give off to signal their readiness to mate. Sorry, these ersatz pheromones coming from flowers won't help a Cassanova. These scents are directed at insects, because insects are the creatures that the flowers want to attract to perform pollination. Still, in old Persia the delicious perfume of *Elaeagnus* blossoms was said to have a powerful effect on a woman's emotions, so much so that husbands reputedly locked up their wives while the trees were in bloom.

Of course, more of a draw than merely scent is needed to keep an insect on a flower. So the mirror orchid deceives male bees that pollinate it by not only smelling like a receptive female, but also by looking like one.

Aroma of amorphophallus is that of rotting flesh.

The attractiveness of the ersatz pheromone is made all the more apparent by the ready suitors that will cluster around a blossom, even if it is experimentally hidden from view by a cloth covering.

After one or two flowers, any smart bee is going to get the hint that he's not dealing with a real female, and give up his efforts. To keep up the deception, each mirror orchid plant does not smell exactly alike, so it takes a half dozen or so plants before a bee catches on, and by then the flowers have got what they wanted. Not so the bees; Nature can be cruel!

Even more intricate in its deception is the bucket orchid of Central America, which splays out little "buckets" filled with a quarter inch depth of perfumed liquid. Each of the 20 or so species of this plant fills its bucket with a scent that varies slightly—although the differences are undetectable to us humans—according to the species of iridescent bee it means to attract. In the flurry of activity around a bucket, an occasional bee falls in. As the bee squeezes out through a narrow tube, the only route out of the bucket, it incidentally pollinates the orchid flower.

I mentioned earlier that floral scents often mimic insect pheromones, but perhaps something more than just pleasant aroma in some floral scents is what makes them appealing to us humans. After all, so many things in nature smell like each other. Perhaps some insect pheromones in those floral scents are also human pheromones, which might explain why we humans have gone to such great efforts trying to capture, to bottle, these aromas.

The first essential oil, attar of rose, was bottled up by the Arabian physician Avicenna about a thousand years ago. Two hundred years later, Eleanore of Aquitaine had twenty-five hundred pounds of violets crushed to make one pound of violet attar, an extravagance even for a queen. That pound of violet attar would soon go rancid, prompting humans to learn to better preserve a scent by combining it with a fixative. Fixatives originally were musk extracted from the genital area of deer and ambergris from sperm whales, but now synthetics are also used.

As an alternative to the elaborate extraction and fixing of floral aromas, I've planted hyacinths, clove currant, and Koreanspice viburnum right outside my back door. There and nearby, a spectrum of scented blossoms provide olfactory thrills from late winter to late fall. Among my favorite scents, in addition to those just mentioned, are 'Chambord' carnations, rugosa rose, 'Rose d'Ispahan', 'Carol Mackie' daphne, and, indoors, poet's jasmine, gardenia, and citrus.

The touch here is that felt by the plants

A recent seed order included a request for a seed packet of "sensitive plant," a plant that is both inedible and homely. I want to grow this plant so that it can entertain me and others with its response to being touched (thigmotropism). Part of the joy in growing plants is watching them respond, albeit slowly, to water, fertilizer, and other environmental stimuli. With sensitive plant, response is rapid, at least to touch, which causes quick and temporary collapse of the leaves. The response can be as quick as a tenth of a second, with the signal, initially thought to be an electrical one, coursing through the stems as fast as 2 feet per second, so you can actually watch leaves of a large plant collapse in a wave of motion after one leaf is touched.

This touch response is the result of both an electrical stimulus, much like that generated in animal nerves, and a chemical stimulus. The chemical, known as Ricca's factor after Ubaldo Ricca, who first noted the chemical response in 1916, had been extracted from plants yet was not identified (as gallic acid beta-D-glucopyranosyl-6'-sulfate) until 1981. In sensitive plant, more recent evidence suggests that the sequential response could also be from changes in hydraulic pressure moving through the xylem.

Response to touch is not uncommon in the plant kingdom. In late April, out in my garden, pea tendrils are pulling my pea vines up a chicken wire fence. A month or so later, bean stems flopping loosely in the air will close in tight spirals on encountering the poles I have set out for them, and clematis leaves, which act like tendrils, are reaching for a wire or twig around which to wrap.

Clematis leaves act like tendrils, hugging support.

And how about the Venus flytrap, which clamps shut its hinged leaf around an unwitting insect?

Plants are discriminating about what they will move for. Venus flytrap can distinguish somewhat between living and dead prey by closing only if two different sensing hairs within its "jaws" are touched in succession, or one hair is touched twice. Tendrils can distinguish between different types of surfaces, responding more quickly to rough or textured surfaces than to smooth or soft surfaces. (The practical implications in the garden are obvious.) Some plants bend toward what touches them, other plants bend away. Beans are less discreet; no matter where they are touched, they move in the same direction, clockwise or counterclockwise, depending on the type of bean.

I wrote that touch response in plants "is not really uncommon." Touch response also includes the stockiness that results when plants are repeatedly touched (or shaken or bruised), mentioned back in the chapter about propagation. Scientists have found that even repeatedly holding a ruler against a leaf to take measurements for experiments influenced leaf size.

Tendrils and twining stems obviously are useful for getting plants up off the ground. Leaves that close around a fly help nourish the Venus flytrap. And stocky growth in response to shaking makes a plant at a windy site better able to withstand the force of wind.

But what use could "fainting" be to a sensitive plant? Rapid collapse of the leaves could help the plant conserve water in drying winds. Or, if a large animal nibbled even one leaf, the plant's lushness would evanesce, leaving an unappetizing skeleton of stems. Rapid collapse might also startle, then scare away, a hungry insect. Perhaps the reason for thigmotropism in this plant is to be found in the plant's scientific name—*Mimosa pudica*. *Pudica* is the Latin word for bashful. Sensitive plant is an entertaining plant, but one that I will not touch too much in an outward show of affection, for fear of collapsing it.

And finally, the efforts I take to grow the best tasting fruits and vegetables

Home grown. Farm fresh. Mmmmm. What could taste better? Vegetables and fruits moments away from being plucked from branches, dug from the ground, or cut from stems are definitely fresher than those merely lifted from market shelves—but they're not necessarily more flavorful. Flavor reflects a melding of a plant's genotype (its genetics) and growing conditions, the latter spelled out by the plant's environment above and below ground.

So what's a gardener to do to be able to harvest the most flavorful carrot, apple, or tomato?

A Measure of Flavor

It's in the nature of humans to quantify and to name. This has been done with flavor, throwing sight, smell, and touch into the melting pot to describe a food in terms of its "organoleptic" properties. Dr. Roger Way, longtime apple breeder for Cornell University, once bemoaned to me Americans' propensities to "eat with their eyes" (back around 1980, at least). His 'Jonagold' apple, a variety he created that combines the sprightly flavor of 'Jonathan' with the honey sweetness of 'Golden Delicious', has a red-splashed, yellow skin, that was no match for the almost flavorless, but shiny red 'Red Delicious' apples then dominating American markets.

The "bells and whistles" of science have been pressed into service in trying to quantify organoleptic qualities (organolepticness?). Soluble solids, which are mostly dissolved sugars, seasoned with organic and amino acids, soluble pectins, and other flavor components, are easily measured with a hand-held refractometer. More telling is to separate individual components in a chromatographic column that selectively adsorbs then desorbs components for identification. In one study, "Commercial '[Red] Delicious' apple essence was extracted to yield an oil with a strong apple-like aroma. The oil was separated into its components with...gas chromatographic columns. Fifty-six compounds were identified." Is 'Red Delicious' really that flavorful!? Instrumental analysis often fails to jive with my favorite method for evaluating quality—tasting.

Flavor Influence: Climate, Microclimate, Water

Knowing how to measure good eating, or just tasting the food, we can look into what influences it. Weather is one influence, a mishmash of rainfall, humidity, and temperature all having their effects, individually and in combination, as well as interacting with soil, genotype and, perhaps, sun intensity and photoperiod. 'Ellison's Orange' is a tasty apple with a delicate hint of anise, but, organoleptically, it falls short for me. This apple originated in the cool summer climate of Britain but becomes too soft in flavor and texture to enjoy when ripened during the hot summer days and nights of my Hudson Valley garden. Experiments have shown that just a few degrees of difference in night temperatures influences both flavor and sweetness of strawberry fruits. The influence of climate, the more general weather pattern, on grape flavor, especially as it relates to wine making, is well known.

The environment is what it is; or is it? Mark Twain wrote, "Everybody complains about the weather, but nobody does anything about it." Not totally so, for us gardeners. Microclimates, pockets where the climate differs from the more general climate, as discussed back in the chapter on "Stress," not only influence plant growth, but also might be sought out in order to eke out the best flavor from a fruit or vegetable. The place to site a fig tree (the fruit of which suffers in flavor under cool ripening conditions—

more or less, depending on variety) in a marginally warm climate would be on a south-facing slope, which captures more incident rays from the sun, or near a south facing wall or fence, which reflects as well as absorbs heat, later to re-radiate it.

I inadvertently create mini-microclimates when I enclose individual apples or bunches of grapes in bags (a commercial practice in Japan!) with the goal of keeping insects, diseases, and birds at bay. The resulting microclimate may affect flavor. Experiments have shown that bagging, besides speeding up ripening, increases flavor aromatics, probably due to increased temperatures and increased buildup of ethylene, a ripening hormone.

Effects probably vary with the color, translucency, and material of the bag, perhaps accounting for why one study showed decreased acidity and increased flavor in bagged apples while another showed the opposite effect with peaches! Bags that block

Bagged grapes protected from birds, insects, and diseases.

Bagging results in near-perfect bunches of grapes.

sunlight allegedly make for less flavorful grapes but, after decades of enclosing grape bunches in white delicatessen bags, I detect no influence on flavor between bunches bagged or left unbagged. (Some light does penetrate the white bags.) My unbagged grapes aren't sprayed with pesticides, so do suffer some affronts from insects and diseases; a better comparison would be between bagged grapes and unbagged grapes, with pests kept at bay in both cases.

Water is one aspect of "weather" over which we have some control—to provide it, at least. Generally, fruiting plants kept a little thirsty bear fruits with better flavor. They are more concentrated in soluble solids and, often, flavor aromatics. Keeping tomato, strawberry, peach, and apple plants slightly thirsty should improve their flavor, at some sacrifice to fruit size. Experiments with grapes demonstrated no effect on flavor, a finding that perhaps could be extrapolated to other fruit plants whose individual fruits are relatively small compared to the size of the plants.

Flavor Influence: Light and Energy

Light is an important part of a plant's environment. It drives photosynthesis which, in turn, creates sugars and other flavor components. With the exception of photoperiod, light is easily amenable to our manipulation—by pruning. Training systems and maintenance pruning of fruit plants are geared to letting all parts of the plant bathe in sunlight. Hence, the central leader system to which I train my apple and pear trees: one central stem and increasingly wide spreading and adequately spaced main stems, going from upper to lower parts of the tree, prevent self-shading. Peach trees naturally have a more spreading growth habit, so I train my tree to the open center system. With three main branches pointing upward and outward from atop a short trunk, the open center system likewise prevents self-shading.

Even well-trained, mature plants will, over time, shade themselves if left to their own devices. The interior of an apple tree becomes leafless limbs bereft of fruit; fruits will be found high up in the sunny periphery of branches. An unpruned grapevine soon becomes a tangled mess. Even an untended strawberry patch eventually gets so overcrowded with strawberry plants that yield, size, and flavor of berries suffers.

(Not as convenient as good pruning but also having a positive impact on flavor is increasing light by laying reflective foil on the ground beneath a plant.)

Flavor might even be affected by the color of light, something usually not practical to fiddle with. Red mulch has been shown to increase the size and flavor of strawberries, probably due to reflection of more red and far-red light inducing changes through phytochrome-mediated regulation pathways. Turnips grown with blue, green, or white mulch produced turnips rated as having "sharp," "mild," or "less distinct" flavor, respectively. Not that any of this is practical, attractive, or integrates well with good soil management.

Weather and the environment, along with genetics, are not the end-all when it comes to flavor. Good flavor takes energy, and a plant has just so much energy. With fruit plants, diverting some of that energy into fewer fruits should pack that much more energy—and flavor—into those left hanging; and it does.

One of the benefits of dormant pruning is that it reduces the number of potential fruits by removing stems that would have borne fruit. Fruit thinning, discussed back in the "Flowering and Fruiting" chapter, further boosts flavors of large fruits by further concentrating the plants' energy into fewer sinks, i.e., fruits. Among small fruits, grape flavor is enhanced not only by dormant pruning but also by pruning off the stem holding the bottom quarter of the berries in the cluster or by pruning off some individual clusters.

Flavor Influence: Harvest Timing

Which brings me to harvest, the timing of which is crucial for eking out peak flavor from any vegetable or fruit. For vegetables, except some fruiting ones such as tomato, pepper, and winter squash, youngsters are most crisp and flavorful. For fruits, best flavor is achieved at peak ripeness, after which senescence sets in and flavor declines. Which is why fruit picked fully ripe and then put into storage declines in flavor.

Some fruits can ripen off the plant if harvested when sufficiently mature. Most have better flavor the closer to full ripeness the fruit is picked. Other fruits, such as avocado, medlar, and European pear, must be picked before fully ripe for ripening off the plant.

Flavor Influence: Soil

Let's now look underground. Confidence in exerting control in this arena was given a boost when the great German chemist Justus von Liebig, in his 19th-century *Die organische Chemie in ihrer Anwendung auf Agricultur und Physiologie* (*Organic Chemistry in Its Application to Agriculture and Physiology*), parsed plants' nutritional needs into simple chemistry. That simple chemistry, although subsequently shown to be only a part of the story, spawned countless experiments spelling out how much plants might need of basic elements.

Most agricultural experiments and recommendations were and are geared to optimizing yield, not flavor (or pest resistance, storability, and other traits). Generally, such sleuthing, when it comes to flavor, tends to be reductive, falling flat on its face when confronted with something for which there are myriad interacting influences. With three exceptions.

First is the case of fertilization. More robust growth, coaxed mostly by adequate nitrogen and water, makes for more succulent and tender leafy vegetables with, perhaps, also better flavor.

Exception two relates nitrogen to the flavor of fruits. Generally, less nitrogen, to a point, leads to better flavor. Less nitrogen also generally leads to lower yields and smaller fruits, so this benefit might trace to mere concentration of flavor, in the same way that keeping plants a little thirsty concentrates flavor. The leaf, wood chip, or hay mulches I spread beneath my mature fruit trees keeps these trees sufficiently well nourished (in addition to the other benefits of these organic materials) as they decompose and release nitrogen slowly and naturally. Where more nitrogen is needed, such as for a young tree, I'll use compost, or compost plus soybean or alfalfa meal, as a more nitrogen-rich mulch.

Exception three applies specifically to onions. Onion flavor depends, first, on genetics. "Sweet" onions, such as the famed 'Vidalia' and other "European" types, are naturally high in water and clock in at more than 6% sugars, while remaining low in pungency. Storage onions, which are the firm, pungent, "American" types, typically have 3 to 5% sugars. So, the varieties to grow, when a sweet onion is wanted, are the various 'Grano' and 'Granex' hybrids, 'Yellow Bermuda', 'Walla Walla', 'Sweet Imperial', 'Candy', 'Sweet Spanish', and 'Nu-Mex Sweet'. (A chosen variety also needs to be adapted to the photoperiod of the region in which it is grown in order to get good bulbing.)

Onion flavor, specifically pungency, also depends on sulfur compounds, which are released when an onion is sliced, chewed, or otherwise damaged. Pungency can be quantified by measuring pyruvic acid, which is a byproduct of the reaction that occurs when onions are sliced or their cells are otherwise damaged. The above mentioned sweet onions clock in at less than 5% pyruvic acid; American types at 10 to 13%.

(When an onion is cut, S-alk(en)yl-L-cysteine sulphoxides, the flavor precursor compounds, are hydrolyzed by the enzyme alliinase to produce thiopropanal S-oxide, which causes tearing, pyruvic acid, and ammonia.)

The amount of sulfur in an onion is tied to the amount of sulfur in the soil. More sulfur, more pungency. For the sweetest onions, sulfur in its elemental form or incidentally with sulfate

'Ailsa Craig' — a genetically sweet onion.

fertilizers such as gypsum, Epsom salts, or ammonium sulfate should not be added to a soil. The sweetest of all onions will be varieties grown on low sulfur soils—conditions which exist in and around Vidalia, Georgia, home, by law, of famously sweet 'Vidalia' onions. Not much can be done if a soil is naturally high in sulfur beyond growing a sweet type and hoping the resulting bulbs are sufficiently sweet. I've never tested my soil for sulfur, but I grow 'Ailsa Craig' and 'Sweet Spanish' onions, and they're sweet enough for me. To still have fresh onions—fresh from storage in my basement—in March, I also grow 'New York Early' and 'Copra'.

As with most biological systems, simplicity does not rule. High temperatures or water stress increase pungency, as does time in storage. But sweet onions are anyway not noted for long storage times. And sometimes I prefer the bite of a pungent onion.

Aggregate Influences

With light, moisture, temperature, day length—so many variables—making their mark on flavor, a more additive approach to growing flavorful crops might be more useful rather than trying to parse out individual, interacting, influences.

This kind of attention has been lavished on studies with carrots by raising them in phytotrons, where light duration and intensity, day and night temperatures, and humidity can be manipulated, in pots of various types of soil. Testing soils and growing conditions mimicking those of Wisconsin, California, Florida, and Texas, the best flavored carrots...drum roll...were those grown in mineral, especially loam, soils as compared with muck soils (drained swamplands rich in organic matter) under mild winter conditions (such as in California). I'm not ready to relocate to be able to grow the most perfect carrot, and no need. Variety choice was still the most important determinant of flavor.

Similar studies have been done with peppers, focusing specifically on their hotness, which, to muddy the waters, stems from not one, but from a whole group of compounds, capsaicinoids, mostly capsaicin and dihydrocapsain. Hotness in peppers was found to depend on the variety, the environment, and the interaction between variety and environment, with smaller fruited peppers less influenced by vagaries of the environment. Usually, but not always, a pepper will have more bite if plants are grown with warmer nights, with colder days, with just a little too much or too little water, or with fertility imbalances; increased elevation elicits the fiercest bite. Basically, with any sort of stress.

So what's a pepperophile, or more specifically, a capsinicoidophile, to do? My tack is to try a few varieties under annually consistent conditions, then continue to grow what does best.

We haven't made much progress in pinpointing how to eke out the best flavor from an apple tree or a grape or tomato vine by micromanaging specifics of growing conditions. Perhaps accounting for the combined effects of environment and variety is still too limiting, singling out too few variables; an even more additive approach might be to compare flavors of organically grown with conventionally grown vegetables and fruits. Which tastes better?

"Organically grown" encompasses a wide range of practices, including choice of variety, soil management, and pest control. To further complicate matters, a good conventional gardener or farmer puts into play many of the same practices as a good organic gardener or farmer. Differences between organically and conventionally grown produce could be in the mineral nutrition of the plants, most notably nitrogen, often lower in organic produce, which is associated with, as mentioned previously, lower yields of smaller, yet more flavorful, fruits. Higher levels of soil organic matter—which should be the case for an organic garden or farm—would affect soil moisture and the spectrum and kinds of soil microbes and nutrients, all with possible flavor effects.

Even pesticide sprays can affect flavor. For instance, sterol-inhibiting fungicides decrease alcohols, aldehydes, esters, and sesquiterpenes—all of which are flavor components—of apple fruits. On the other hand, emulsions of soybean oil sprayed three weeks and a few days before harvest of 'Golden Delicious' apples resulted in better flavor at harvest and after storage.

Again, findings vary. Some studies have found that people prefer the flavor of organic produce such as strawberries, tomatoes, and apples, but no broader, consistent flavor differences have been found. How about this? Animal tasters, in one study, preferred organic—which may or not be true for different animals than those "volunteered" for the study.

Terroir

In desperation, we may need to fall back on terroir, a concept that takes into account almost everything: soil, climate, slope, sunlight, the geology and geography of a particular region, and as much more as needed to lend it a certain mystical quality. Historically, terroir has been associated with wine, tying the quality of a wine to the terroir rather than to the vintner.

There's an appealing *je ne sais quoi* to terroir—that therein lies something beyond what science could possibly quantify. "Wines express their source with exquisite definition. They allow us to eavesdrop on the murmurings of the earth," wrote wine critic Matt Kramer (*Making Sense of Wine*). Of a California vineyard's highly regarded chardonnays, he writes, there is "a powerful flavor of the soil: the limestone speaks."

Why not broaden the scope of terroir, applying it to other fruits and vegetables? Perhaps it is a certain terroir that brings out the best flavor of a tomato, or one that brings out the best flavored fruits from a peach tree, and so on.

We love to love terroir but, sad to say, studies have mostly debunked the notion. Terroir and the legal designations it engenders are, yes, tied to the price of a bottle of wine or the price of vineyard land—but, no, not to the flavor or quality of the wine. The flavor of fruits and vegetables—and wine, no doubt—are influenced by soil, environment, light, and the myriad ways each of these influences can vary and interact. The relationships might not yet be thoroughly understood, but that's no reason to sit back and pin flavor on some indefinable quality of "place."

The upshot of all this is that terroir and organically grown may affect flavor—or not. They most assuredly do affect perception of flavor and, hence, market returns. As the eternal skeptic, I am ready to debunk a concept like terroir, less so organic (the flavor effect, not the practice).

Still, I can't help but add my own "anecdata," which has dulled my natural skepticism. One variety of apple that I grow is 'Macoun'. When I harvest them at the right moment—which is just as soon as they hit the ground—their flavor is *ne plus ultra*. Five miles away, I've picked the same variety from a commercial orchard, harvesting fruits from the ground as well as those that yield to the slightest twist and upward lift.

My 'Macoun' apples taste very different, and much, much better than those from the nearby commercial orchard. The sites are very different, mine being low lying with deep, clay loam soil, in contrast to the clayey, rocky hillside of the commercial orchard. Their site has better air; I have what would generally be considered to be a naturally better soil. I've further improved the soil beneath my trees by making it rich in organic matter, the result of compost at planting time and yearly additions of hay, wood chip, and leaf mulches. The commercial orchard gets herbicide, synthetic fertilizers, and repeated sprays of various pesticides throughout the growing season. Rootstocks also differ between sites.

All this, is it terroir? Is it the organic vs. conventional treatment? To throw one more wrench into the works, I wouldn't discount the halo effect: I, like almost everyone, enjoy most the flavor of the apple that I grew. Does it really taste better?

Enough waffling. Can an understanding and application of a little natural science make for tastier fruits and vegetables? To some degree, with some fruits and vegetables, yes. To sum up...

- Genetics is very important, usually the most important determinant of flavor.
- Climate is important for some fruits and vegetables, such as carrots.
- Soil, to a lesser extent, can be important, although what about the soil is important for a specific crop is hard to pinpoint.

As we gain deeper understanding of the natural science behind the scenes in a garden, influences on flavor will, no doubt, be more quantifiable.

For now, my tack is to choose the best-tasting varieties for my region (see "Organizations" and "Propagation and Planting" chapters); to site and to plant according to the weather and the climate (see "Stress" chapter); to prune and to regulate production (see "Flowering and Fruiting" and "Stems and Leaves" chapters); and to provide good growing conditions below ground, mostly through frequent applications of an abundance of organic materials (see "Soil" chapter). All this makes for very good tasting vegetables, fruits, and herbs.

EPILOGUE: THE SCIENTIFIC METHOD

Charles Darwin did some of his best work lying on his belly in a grassy meadow. Not daydreaming, but closely observing the lives and work of earthworms, eventually leading to the publication of his final book, *The Formation of Vegetable Mould through the Action of Worms*. He calculated that these (to some humans) lowly creatures brought 18 tons of nutrient-rich castings to the surface per acre per year, in so doing tilling and aerating the soil while rendering the nutrients more accessible for plant use. He wrote that "worms played a more important role in the history of the world than most persons would at first suppose."

We also can take a more scientific perspective in our gardens without the need for digital readouts, flashing LEDs, spiraling coils of copper tubing, or other bells and whistles of modern science. What's most needed is careful observation, an eye out for serendipity, and some objectivity.

Observation invites questions. How many tons of castings would Darwin's earthworms have brought to the surface of the ground in a different soil? Or from soil beneath a forest of trees rather than a grassy meadow?

And questions invite hypotheses, based on what was observed and what is known. Darwin's prone observations, along with knowledge of soils, earthworms, plants, and climate, might invite a hypothesis such as "Earthworms would bring a greater amount of castings to the surface in a warmer climate." Is this true? How can we find out?

* * *

Gardens are variable and complex ecosystems, which makes growing plants both interesting and, if you want to know why a plant did what it did, frustrating. Many gardeners do something—adding Epsom salts to the soil for better growth of tomatoes, for example—and attribute whatever happens in the ensuing season to the Epsom salts, ignoring the something else, or combination of things, that might also have made contributions to whatever happened.

Enter the scientific method, a way to test a hypothesis. You put together a hypothesis by drawing on what is known and what can be surmised. Say you have two tomato plants and you spray one with a "tea" made by steeping nettle leaves in water. Based on "what is known and what can be surmised," stinging nettle spray could, in fact, nudge along tomato growth. After all, nettle leaves contain nutrients, some of which would steep into the "tea." Plants can take up some nutrients through their leaves. And many gardeners do, in fact, recommend nettle tea to boost plant growth.

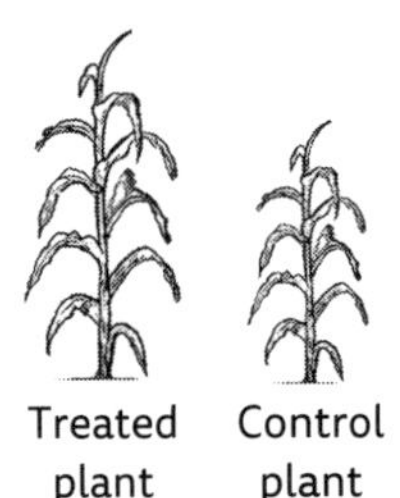

Your sprayed plant growing better than your other, unsprayed plant would strengthen the case for further study. Why further study? Because the response of two plants in a given season at a given location is not sufficient to make a general recommendation.

The way to truly assess the benefit of the spray is to subject it to scientific scrutiny: Come up with a hypothesis, such as "Nettle spray makes tomato plants grow more" (or yield more, or have less disease, or whatever other hypothesis is being tested), and then design an experiment to accurately test the validity of the hypothesis.

* * *

A well-designed experiment would need more than just one treated (nettle sprayed) plant and one control (water sprayed) plant. Grow ten tomato plants of the same variety under the same conditions and some will grow a little more than the others, some a little less. With too few test plants, natural variation in growth from plant to plant might overwhelm any variation due to a treatment (spraying with nettles in this example). Given enough plants to even out the natural variations in, say, plant growth, the effects of a treatment can be parsed out. Greater natural variations would require more plants for the test.

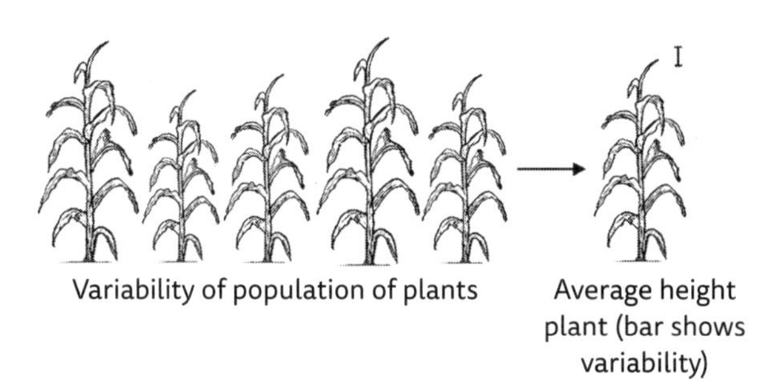

A garden experiment might have additional sources of variation. Perhaps one side of a plot is more windy, or the soil is slightly different, or basks plants in a bit more sunlight than the other side. Rather than have all the treated plants cozied together growing better or worse because of this added effect, even out these effects by randomizing the locations of treated and control plants.

Now we've got an experiment! All that's needed is to spray designated plants with either the nettle tea or the water, and then take measurements. Plug those measurements into a software program for statistical analysis and a computer will spew out a

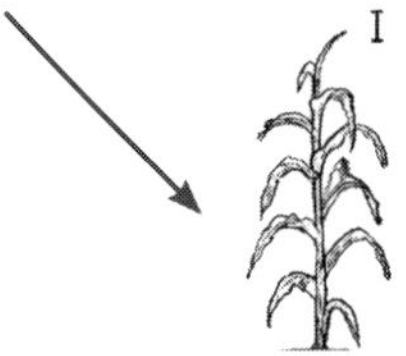

percent probability, based on variability within and between each group of plants, that the tea was responsible for increased plant growth. In agriculture, a test with 90% or 95% probability is usually considered sufficient to link cause and effect. You can then answer "yea" or "nay" to the hypothesized question—in this example, "Does a nettle spray make tomatoes grow better?"

* * *

A good test could involve a lot of plants and a lot of measurements, more than most of us gardeners are willing to endure. A danger exists, as Charles Dudley Warner so aptly put it in his 1870 book, *My Summer in a Garden*: "I have seen gardens which were all experiment, given over to every new thing, and which produced little or nothing to the owners, except the pleasure of expectation." Then again, setting up something less than a full-blown experiment could be fun and, while not proving something to a 95% confidence level, still suggest a possible benefit.

Knowing what's involved in testing a hypothesis offers appreciation for all that can affect plants. Perhaps the good growth of the tomato plant wasn't from your nettle spray. Knowing something of the scientific method can help you assess, whether observed in your own garden or a friend's garden, or reported in a scientific journal, the benefit of the spray.

So, fellow gardener, go out to your garden and look more deeply into Nature, perhaps, like Darwin, lying on your belly. Understanding some of the science at play in the garden takes it to the next level. And you'll find that the real world, neatly woven together, is imbued with its own poetry, with science being one window into that poetry.

Index

About the Author

Credit: Vicki Herzfeld Arlein

LEE REICH, PhD, is an avid farmdener (more than a gardener, less than a farmer) with graduate degrees in soil science and horticulture. After a few years in agricultural research with the USDA and then Cornell University, he went off on his own, writing, lecturing, and consulting. His writing includes a number of gardening books and a bimonthly column for Associated Press. On his farmden, in New Paltz, NY, he hosts workshops and training, which helps satisfy his self-imposed educational mandate. The farmden is also a test site for innovative techniques in soil care, pruning, and growing fruits and vegetables. If he is not in the garden, he might be found learning to play jazz, blues, and latin rhythms on his ukulele, or fashioning wood into tables, cabinets, doors, and other objects useful and beautiful, sometimes using wood of an old or no longer productive tree he once planted.

ABOUT NEW SOCIETY PUBLISHERS

New Society Publishers is an activist, solutions-oriented publisher focused on publishing books for a world of change. Our books offer tips, tools, and insights from leading experts in sustainable building, homesteading, climate change, environment, conscientious commerce, renewable energy, and more—positive solutions for troubled times.

DON'T EAT THIS BOOK *(but you could)*

We're proud to hold to the highest environmental and social standards of any publisher in North America. This is why some of our books might cost a little more. We think it's worth it!

- We print all our books in North America, never overseas
- All our books are printed on 100% **post-consumer recycled paper**, processed chlorine-free, with low-VOC vegetable-based inks (since 2002)
- Our corporate structure is an innovative employee shareholder agreement, so we're one-third employee-owned (since 2015)
- We're carbon-neutral (since 2006)
- We're certified as a B Corporation (since 2016)

At New Society Publishers, we care deeply about *what* we publish—but also about *how* we do business.

New Society Publishers

ENVIRONMENTAL BENEFITS STATEMENT

For every 5,000 books printed, New Society saves the following resources:[1]

23	Trees
2050	Pounds of Solid Waste
2255	Gallons of Water
2941	Kilowatt Hours of Electricity
3726	Pounds of Greenhouse Gases
16	Pounds of HAPs, VOCs, and AOX Combined
6	Cubic Yards of Landfill Space

[1] Environmental benefits are calculated based on research done by the Environmental Defense Fund and other members of the Paper Task Force who study the environmental impacts of the paper industry.